W0235142

▶ **On Face Transplantation**

DOI: 10.1057/9781137452726.0001

Other Palgrave Pivot titles

Graham Oppy: **Reinventing Philosophy of Religion: An Opinionated Introduction**

Ian I. Mitroff and Can M. Alpaslan: **The Crisis-Prone Society: A Brief Guide to Managing the Beliefs That Drive Risk in Business**

Takis S. Pappas: **Populism and Crisis Politics in Greece**

G. Douglas Atkins: **T.S. Eliot and the Fulfillment of Christian Poeticss**

Guri Tyldum and Lisa G. Johnston (editors): **Applying Respondent Driven Sampling to Migrant Populations: Lessons from the Field**

Shoon Murray: **The Terror Authorization: The History and Politics of the 2001 AUMF**

Irene Zempi and Neil Chakraborti: **Islamophobia, Victimisation and the Veil**

Marian, Duggan and Vicky Heap: **Administrating Victimization: The Politics of Anti-Social Behaviour and Hate Crime Policy**

Pamela J. Stewart and Andrew J. Strathern: **Working in the Field: Anthropological Experiences across the World**

Audrey Foster Gwendolyn: **Hoarders, Doomsday Preppers, and the Culture of Apocalypse**

Sue Ellen Henry: **Children's Bodies in Schools: Corporeal Performances of Social Class**

Max J. Skidmore: **Maligned Presidents: The Late 19th Century**

Lynée Lewis Gaillet and Letizia Guglielmo: **Scholarly Publication in a Changing Academic Landscape**

Owen Anderson: **Reason and Faith at Early Princeton: Piety and the Knowledge of God**

Mark L. Robinson: **Marketing Big Oil: Brand Lessons from the World's Largest Companies**

Nicholas Robinette: **Realism, Form and the Postcolonial Novel**

Bernadette Andreosso-O'Callaghan, Jacques Jaussaud, and Maria Bruna Zolin (editors): **Economic Integration in Asia: Towards the Delineation of a Sustainable Path**

Umut Özkırımlı: **The Making of a Protest Movement in Turkey: #occupygezi**

Ilan Bijaoui: **The Economic Reconciliation Process: Middle Eastern Populations in Conflict**

Leandro Rodriguez Medina: **The Circulation of European Knowledge: Niklas Luhmann in the Hispanic Americas**

Terje Rasmussen: **Personal Media and Everyday Life: A Networked Lifeworld**

Nikolay Anguelov: **Policy and Political Theory in Trade Practices: Multinational Corporations and Global Governments**

Sirpa Salenius: **Rose Elizabeth Cleveland: First Lady and Literary Scholar**

Sten Vikner and Eva Engels: **Scandinavian Object Shift and Optimality Theory**

Chris Rumford: **Cosmopolitan Borders**

Majid Yar: **The Cultural Imaginary of the Internet: Virtual Utopias and Dystopias**

DOI: 10.1057/9781137452726.0001

palgrave▸pivot

On Face Transplantation: Life and Ethics in Experimental Biomedicine

Samuel Taylor-Alexander

The University of Auckland, New Zealand

palgrave
macmillan

DOI: 10.1057/9781137452726.0001

© Samuel Taylor-Alexander 2014

All rights reserved. No reproduction, copy or transmission of this publication may be made without written permission.

No portion of this publication may be reproduced, copied or transmitted save with written permission or in accordance with the provisions of the Copyright, Designs and Patents Act 1988, or under the terms of any licence permitting limited copying issued by the Copyright Licensing Agency, Saffron House, 6–10 Kirby Street, London EC1N 8TS.

Any person who does any unauthorized act in relation to this publication may be liable to criminal prosecution and civil claims for damages.

The author has asserted his right to be identified as the author of this work in accordance with the Copyright, Designs and Patents Act 1988.

First published 2014 by
PALGRAVE MACMILLAN

Palgrave Macmillan in the UK is an imprint of Macmillan Publishers Limited, registered in England, company number 785998, of Houndmills, Basingstoke, Hampshire RG21 6XS.

Palgrave Macmillan in the US is a division of St Martin's Press LLC, 175 Fifth Avenue, New York, NY 10010.

Palgrave Macmillan is the global academic imprint of the above companies and has companies and representatives throughout the world.

Palgrave® and Macmillan® are registered trademarks in the United States, the United Kingdom, Europe and other countries.

ISBN: 978–1–137–45273–3 EPUB
ISBN: 978–1–137–45272–6 PDF
ISBN: 978–1–137–45271–9 Hardback

Library of Congress Cataloging-in-Publication Data

Taylor-Alexander, Samuel, 1984–
    On face transplantation: life and ethics in experimental biomedicine/Samuel Taylor-Alexander, the University of Auckland, New Zealand.
        pages cm
        ISBN 978–1–137–45271–9 (hardback)
            1. Face – Transplantation. 2. Surgery, Plastic – Moral and ethical aspects. 3. Biotechnology. I. Title.

RD523.T39 2014
617.5′2059—dc23   2014028154

A catalogue record for this book is available from the British Library.

First edition: 2014

www.palgrave.com/pivot

DOI: 10.1057/9781137452726

In memory of Frank and Nora Alexander

DOI: 10.1057/9781137452726.0001

Contents

DOI: 10.1057/9781137452726.0001

Acknowledgements

Thank you to everyone in Mexico whose openness allowed me to tell this story, especially Raymundo Priego Blancas, Jose Haddas Tame, and Octavio Madrid Mata. In writing on this important topic I have benefited greatly from the mentorship and kindness of Sheila Jasanoff and Monique Skidmore, from the friendship, guidance and encouragement of Susanna Trnka, Julie Park, Catherine Trundle, and Ruth Fitzgerald, and from the love of Courtney Addison.

Those close to me know how tumultuous things have been over the last few years. Siblings Emile, Patrick, and Sian continue to be a source of comfort and joy. To my parents, thank you for everything.

Elements of Part 1 and Part 2 appeared previously in two publications: 'Bioethics in the Making: "Ideal Patients" and the Beginnings of Face Transplant Surgery in Mexico,' *Science as Culture* 23(1), 27–50 and 'On Face Transplantation: Ethical Slippage and Quiet Death in Experimental Biomedicine,' *Anthropology Today* 29(1), 13–16.

palgrave▶**pivot**

www.palgrave.com/pivot

1

A History of the Present

Abstract: *I chart the key institutional, technical, and medical developments that paved the way for the emergence of face transplantation as a therapeutic possibility. These range from developments in surgical technique and immunosuppressive therapy, through to new ways of categorizing people whose hearts continue to beat while they lay in a state of 'irreversible coma.' In doing so, the I lay out the main issue that medical teams have faced in their pursuit to transform face transplant surgery from a medical possibility into a clinical reality: How can doctors expose a person with a non-life threatening condition to a potentially fatal treatment regime in order to improve their quality of life? In negotiating the ethical and technical challenges related to the operation, plastic surgeons around the world began to categorize patient groups, delineating in the process an ideal face transplant patient. This is a patient, they argue, who is healthy enough to withstand the stresses of the operation and ill enough to make the transplant a clinical necessity.*

Taylor-Alexander, Samuel. *On Face Transplantation: Life and Ethics in Experimental Biomedicine.* Basingstoke: Palgrave Macmillan, 2014. DOI: 10.1057/9781137452726.0003.

Medicine on the make

A patient is hooked up to a respirator, lying in a state of irreversible coma on a bed at Brigham and Woman's hospital. Surgeons shave parts of her head and then use a blue marker to map out incision points. One of them calls for a scalpel and the theatre nurse places it in his right hand. Her heart is still beating as with precise vigour he slices from the top of the patient's forehead, down around her jaw, beneath her chin, and then up to the original incision point. Various instruments are inserted into the area to aid the removal of her face. It is slowly and carefully pulled away from the skull. Another surgeon with electric saw in hand moves in and begins dissecting her jaw. The team now has the requisite elements of face and she is taken off the respirator and allowed to move from a brain dead state to completely dead.

The emergence of face transplant surgery opens up new questions about what it means to be human in the age of high-tech biomedicine. Today, a relatively small number of people have had their faces removed while their hearts, though technologically assisted, continue to beat. By faces I refer to a composite of skin, bone, muscles, fat, jaw and teeth. These grafts are increasingly moved from one human head to another in what medical professionals refer to as the 'allotransplantation of composite facial tissue.' Transplanting composite tissue is nothing new. Doctors, some more successful than others, have been moving kidneys, ears, and other body parts between animals of the same or different species for centuries. Organ transplantation between humans is now a common life saving medical procedure. These operations are dangerous: if the operation doesn't kill you, the drugs you need to avoid rejection will almost certainly harm your body.

Harm is an important issue in biomedicine. It is central to the professional ethic of surgeons and others caring for the sick: first, do no harm. This imperative has been translated into and is secured by different bureaucratic apparatuses such as risk-benefit analysis. It is too easy to examine face transplant surgery in the terms provided by the medical establishment and its external regulators. The various tactics employed by surgical teams around the world and even the joy expressed by face transplant patients and donor families alerts us to this – as does their disgust. As patients delight in their new felt normality, donor families bask in the gift of giving and in seeing loved ones continue to live on almost literally in another person's face.

DOI: 10.1057/9781137452726.0003

Face transplant surgeon Maria Siemionow has spoken about the animals, the rats and rabbits that she practices on, as chimeras, as hybrid specimens. Common organ transplant operations result in patients having the biology of another person pulsating within them, pushing blood through or cleansing the body. Sometimes they reflect on this and give thanks to the donor family. They grapple with how to live with the claims the family has over and into their life, and with having another person's body parts within them. Face transplant patients have someone else's face in place of their own, displaying the donor's mole, brow, or freckle. They have a nose they might struggle to view as their own. They live with joy and sorrow. It is my contention that this hybridity provides a unique opportunity for reflection: What does it mean to be an individual in an age where people can be composed of more than one body? How are understandings of 'life' being reworked with the emergence of new technologies? Are face transplant recipients hybrid beings?

To get at these questions I provide *a* story of face transplantation. Tracing the emergence of new phenomena can break down the taken for granted nature of our present. I draw on a number of different accounts of the growth of organ transplantation, reconstructive surgery, and the like, and supplement these with my own research on face transplantation. This essay is based primarily on interviews with a team of surgeons in Mexico, and on academic and popular accounts of the operation. It shows that face transplantation is something of a hybrid medical practice composed of two sets of techno-medical expertise: organ transplantation and plastic surgery. It is in Mexico where we met our first patient, Escobar.

In the early months of 2007 a team of Mexican reconstructive surgeons became responsible for treating the man I call Escobar. A peasant farmer, Escobar had suffered from an abscessed tooth. Living away from the infrastructure of Mexico's large cities, he first sought but was denied treatment at his local health clinic. Second, he went to a traditional healer who applied a mud lotion to the tooth. The infection did not go away. In fact, it only got worse. It spread, and in doing so took with it large bits of his face: he lost half of his upper lip, half of his lower lip, parts of the floor of his mouth, and almost all of the tissue from his cheek, neck and chin. As well as much of his face he lost forty kilograms of weight, and almost his life. He was interned in one of Mexico City's largest public hospitals.

The severity and particularity of Escobar's condition posed serious problems for the reconstructive surgeons responsible for treating him:

DOI: 10.1057/9781137452726.0003

According to the surgeons I spoke to, restoring function using traditional methods would be 'practically impossible' and so the team began 'working with the idea' of transplanting composite facial tissue. In the emerging and experimental terrain of face transplant surgery, the team was treating 'a patient that we would consider an ideal case.' In other words, the surgeons felt that Escobar's condition not only justified and required, but also obligated the use of face transplant surgery. In order to make sense of the team's opinion I came to explore the various ways in which they situated themselves and were positioned in a specific temporal and bureaucratic context. Below, I use these reflections to pull out broader insights into the past, present, and future of face transplant surgery.

Disciplinary norms

The case of face transplantation demonstrates that a number of ideals surrounding the human-self exist within and alongside medical, scientific, and national imaginaries. Patients are often framed or viewed in relation to the political and epistemic conditions in which they are incorporated, and the various assemblages through which they are produced as objects of care and enquiry. The majority of social scientists that have dedicated substantial analytic time to investigating reconstructive surgery come from the world of psychology, a discipline to which plastic surgery *writ large* owes much of its status as a legitimate form of medicine. While the case studies presented in such works often provide important forays into patient experience, they offer little critical insight into the *a priori* social and political understandings that underscore and structure the practice of plastic surgeons. Rather, they reproduce such understandings as they work from a long institutionalized model of personhood that is based on deeply ingrained concepts of normality *and* that outlines a specific separation and relationship between mind and body.

This can be seen in the early evaluation of face transplantation as a possible medical treatment. As we will see, a number of nationally situated institutional bodies reflected on whether or not the operation should be performed. The review offered by the Working Party of the Royal College of Surgeons of England, which became central to the Mexican case, stated the following, for example: 'The aim of facial transplantation would be to replace unacceptable grafts and flaps with tissue that has the appearance of a normal face and allows mobility of the deeper

DOI: 10.1057/9781137452726.0003

structures.'[1] What constitutes a normal face is unexplored in this report, something that is significant in a context where the face under question is being donated from one person to another. In this context the face under question will look like a mix of donor and recipient, it will be the site of intense reflection by transplant patients as they learn how to eat and speak again, touching the inside of 'someone else's' mouth with their tongue.

A scarce but important body of literature has begun to shed light on how the underlying assumptions that operate in reconstructive surgery shape both the practice and the lives of patients within it. In particular, scholarship from the social sciences and humanities has shown how questions of norms and normality underlie the discipline. Whether manifest in the balancing of sameness and difference by patients and carers, or seen in tactical attempts by affected teens to makes themselves more normal in their own eyes and those of their peers, it is clear that reconstructive surgery is attached to both social and biomedical norms. That is, reconstructive surgery seeks to transform the bodies of patients into an *ideal* prior state based on social and medical understandings of normality.

Another line of critical scholarship emphasizes the historical fluidity of the boundaries between reconstructive and cosmetic plastic surgery. This body of work points the spotlight towards the role plastic surgeons and the medical establishment takes to justify their practice. Here plastic surgery is a place where people engage in 'boundary work' designed to sharpen the blurry lines between aesthetic and reconstructive surgery in order to maintain the legitimacy of the latter as an ethical and necessary practice. Accepting these boundaries plays into rather than questions *a priori* assumptions about normalcy and how we should act with and upon our bodies.[2] Meanwhile particular surgical operations continue to cross boundaries from one side of the discipline to the other – cleft palate surgery was at the end of the 19th century considered a cosmetic proce-dure, for example[3] – and new categories such as 'aesthetic reconstructive surgery' emerge as the contexts of surgical practice become increasingly influenced by the particularities of local health care systems.[4]

The historical accounts of the development of the discipline empha-size the important role that popular and expert understandings of the psychological dimensions of personhood have played in legitimating diverse surgical practices. In noting that all forms of plastic surgery are directed towards the production of personal happiness, for example,

DOI: 10.1057/9781137452726.0003

Sander Gilman[5] also notes that the distinction between mind and body, intrinsic to how we understand ourselves as persons, is central to the credibility of plastic surgery as a medical discipline. The discipline of plastic surgery is based on and incorporates a particular, dualistic notion of personhood in which the mind and body are seen as separate yet mutually constituting aspects of the human self.

Such critical analyses of plastic surgery provide an important starting point for any social enquiry that deals with face transplant surgery. They describe the ethic of reconstructive doctors as one based on social and medical conceptualizations of normality. Medical anthropologists have drawn on a line of French philosophical scholarship that demonstrates that it is by distinguishing the normal from the pathological that biomedicine produces the object of its enquiry. Such demarcation is tied to both to the condition of pathology in biological process and to dominant social and political norms. Moreover, the very idea of the normal and normality has a moral quality in that it denotes what should be and thus becomes something to strive towards.[6] Following this notion we may add that, like other forms of medical practice, reconstructive surgery has an intrinsically political dimension: It seeks to (re)construct people in accordance with institutionalized ideals of personhood.

Surgical beginnings

For reconstructive plastic surgeons, face transplant surgery is viewed as an answer to the limitations of the techniques and the (especially human biological) resources available to them. Every day these doctors perform reconstructive operations that involve moving skin, flesh, and bone around the individual bodies of their patients, often from torso or limbs to their face – they are experts in autotransplantation. While such rearranging of the human body and its parts often produces results that come close to meeting the desires of both patient and practitioner, in other cases surgical constraints are experienced as personal shortcomings, even failures. The desires and hopes of those involved in facial reconstruction have undoubtedly shifted since a decade ago, when serious discussion began regarding the allotransplantation of composite facial tissue – harvesting facial tissue from brain dead donors and transplanting it to restore the anatomy of craniofacial patients.

DOI: 10.1057/9781137452726.0003

In its emergence as a reconstructive option, face transplant surgery has shifted understandings of the efficacy and limitations of common reconstructive techniques, how these limitations are understood, and how they are experienced. Whether or not the desire to perform the operation is another example of how, within an economy of hope, clinicians simultaneously embrace and are embraced by new medical technology,[7] it is more or less accepted that face transplantation offers plastic surgeons the ability to restore function and appearance to a standard not possible with classical reconstructive methods. However, this has not translated into an overall acceptance of the procedure as a treatment option. Amongst other things, the surgery was and still is many kinds of experiment: ethical, technical, legal, social.

Face transplant surgery involves taking a mixture of composite soft and hard tissue from a brain dead donor and transplanting it to a recipient who, in the majority of cases to date, has suffered disfiguration following physical trauma. Like with cases of solid organ and limb transplantation, a strong immunosuppressant regime is needed to counter biological rejection of the transplant. The side effects of immunosuppressant drugs on patients are well documented, and include serious infection and increased possibility of cancer due to a compromised immune system. While these side effects are pertinent issues in the ethical considerations of such operations, the story of organ transplantation begins with physicians who struggled even to move biological tissue around the body of an individual.

The beginnings of face transplant surgery can be traced back to the early 20th century and the work of surgical pioneers who through animal experimentation paved the way for contemporary organ transplantation. Following the anthropologist Lawrence Cohen we can identify three technological shifts that led us toward the related medical present: (1) the development of surgical techniques, such as arterial suture, that solved the problem of how to keep an organ alive when moved to a different part of the subjects body; (2) the growth of immunological techniques for analysing tissue relatedness that followed the development of transfusion medicine in the Second World War; and (3) the development of immunosuppressant therapy paved the way for the growth and even globalization of transplant surgery.[8]

The first of these shifts took place in the lead up to the 20th century. Physicians from around Europe and the United States came to solve some of the fundamental epistemic and clinical problems that hindered early

DOI: 10.1057/9781137452726.0003

forays into experimental biomedicine and related attempts to remedy illness. Developments in immunology and understandings of organ function emerged especially through efforts to cure goitre, and were followed by attempts to replace human organs through both auto- and xenotransplantation. Clinicians moved between different disciplines (from vascular surgery to transplant experimentation, for example) as the field became the 'order of the day' by the end of the first decade of the 20th century. In doing so, these experimental practitioners brought their expertise to what was a steadily growing discipline. After numerous failed experimental studies, researchers in important medical institutions moved attention from animal-based to human-centred organ transfer. These efforts coincided with emergent understandings that are now fundamental to both general and transplant medicine, especially immunology and organ function. Moreover, surgical and immunological techniques partially solved the problem of biological rejection, making it possible to find a close hematologic match between donor and recipient.[9]

Meanwhile other technical problems remained in the realm of organ transplantation: the only way to procure an organ was by effectively killing the donor. This changed in the aftermath of the Second World War, where an assemblage of historical, institutional, and technical factors converged to create a new category of person: the brain dead. This new way of conceiving of persons who were still physically alive yet *dead enough* to allow for organ procurement coincided with new technologies for preserving life in its barest form.[10] The solidification of the category itself is often traced back to a 1968 report published by an *ad hoc* Committee from Harvard Medical School on what they termed 'irreversible coma.' In this report, the Committee highlighted a profound movement away from traditional conceptions of life 'as judged by the ancient standards of persistent respiration and continuing heartbeat.'[11] In doing so they noted that these characteristics alone are no longer symbolic of life because they don't take into account the complete loss of conscious that would otherwise be present given such biological function.

While tissue typing and the notion of brain-death are often attributed as the core developments in the history of organ transplantation (see below), the invention of immunosuppressant therapy was also essential in the emergence of the related surgery as a common treatment operation. Though effective in sustaining bodily health in the face of infection or foreign bodies, the immune system becomes an obstacle in the context of transplant surgery. The notion of histocompatibility is central to immunity

DOI: 10.1057/9781137452726.0003

and tissue transplantation, and in basic terms can be explained as follows. Every individual has a host of genetically-coded antigens that sit benignly on the surface of their cells; these antigens are so numerous and diverse that no two individuals have the same antigen profiles (excluding identical twins). When the body detects antigens from outside of itself (for example, if tissue from another body is transplanted), its automatic response is to defend itself: the foreign antigens are presented to B-cells, which produce antibodies to eliminate them. This is how the body keeps itself safe from damaging influences (e.g. disease) but it also prevents foreign (i.e. transplanted) tissue being incorporated into the body. The development of synthetic immunosuppressants circumvents this, disabling the immune system by inhibiting interaction between B-cells.

During the early 1970s, employees at the Swedish pharmaceutical manufacturer Sandoz, later known as Novartis, began screening tests of possible immunosuppressant drugs. One of these substances was cyclosporine, and the discovery of its effects was to become perhaps one of the most important findings in the history of organ transplantation. An incredibly rare form of peptide, the chemical compound was first derived from a fungus, and was later found to interfere with the activity and growth of antigenic cells. Results of successful use in kidney and liver transplant were published before the drug received approval for general use in these and related operations in 1983. Summarizing its import in the history of the growth and globalization of organ transplantation, Lawrence Cohen writes that with cyclosporine, 'close matching of transplant tissue was no longer essential. The game was suddenly not to improve the recognition apparatus but to suppress it altogether.'[12] The required suppression of the recipient's immune system is perhaps the most pertinent ethical issue in the field of face transplantation.

The debate surrounding the transplantation of composite facial tissue has more or less hinged on two issues. Often the matter of primary concern is that the operation entails exposing a person with a non-life threatening condition to a potentially fatal treatment regime in order to improve their *quality of life*. This risk comes primarily through the side effects of immunosuppressant drugs needed to counter rejection of the transplanted graft. A recent summary of the present of the operation noted that, for example:

> Nowadays and, in order to prevent [face transplant] rejection, non-specific induction of overall immunosuppression is used as a triple therapy based on postoperative administration of mycophenolate mofetil, tacrolimus and

DOI: 10.1057/9781137452726.0003

steroids. This immunosuppressive regimen should be continued throughout life, which leads to toxicity risks, infectious complications (opportunistic infections with cytomegalovirus, herpes, etc.), metabolic complications (diabetes), nephrotoxicity, hypertension and tumors.[13]

In the case of those now common solid organ transplant procedures (of hearts, kidneys, livers), such risks have been accepted because the operations are deemed (biologically) life saving.

The second issue relates to a deficit in knowledge regarding the risks and benefits of the operation. Debate on the matter has resulted in new arguments about the very way that the human body and mind function and respond to different substances, human and non-human. Initial arguments focusing on immunology have centred on both the efficacy of immunosuppression for countering the antigenic property of the skin, which is often held to be stronger than other organs, and on the potential side effects of immunosuppressant drugs: possible infection leading to death. Others, including those made by prominent bioethicists, have questioned the affective dimensions of the operation, especially noting the unknown dimensions of living with someone else's facial flesh in place of your own. Many of these arguments have drawn on examples from hand transplantation, and the notable case of 'psychological rejection' when a patient was unable to bear living with another person's hands in place if his own, and asked for them to be removed.[14]

Because of especially the difficulty in calculating the affective dimensions of the surgery – the risks and benefits of living with someone else's flesh in place of your own – plastic surgeons were and continue to be caught in something of a catch-22, a situation where the knowledge they need to perform the operation can only be gained through its performance. This, as well as the mentioning of risks and benefits, alludes to the observation that debate surrounding the surgery has been cast largely in bioethical terms. Attempts to resolve this scenario produced a series of ethical experiments, many of which revolved around the delineation of a patient that could be operated upon given the continuing epistemic and (bio)ethical constraints.

Ideal patients

This coming together and interaction of different technologies, knowledges, modes of reasoning, bureaucratic apparatuses and ethical

DOI: 10.1057/9781137452726.0003

sensibilities has introduced into the imagination of plastic surgeons what my interlocutors often refer to as an 'ideal' face transplant patient. The realization qua justification of the surgery continues to be mediated by a mode of reasoning that suggests that the risks and benefits of the operation can be adequately worked upon through careful patient selection. Early attempts to gain institutional sanction by medical teams in France, the USA, and Britain resulted in a series of published arguments that sought to demonstrate that operating on otherwise physically and psychologically healthy people can pull the risks and benefits of the surgery in favour of its performance for patients suffering severe disfiguration (see below). By way of necessary example, we can move briefly to talk about the one face transplantation that had been performed when the team started developing the protocol.

The first allotransplantation of composite facial tissues was performed in France on 25 November 2005. A team of French surgeons had for a number of months been searching for an ideal patient on whom to perform the operation. They thought that they had found her in Isabelle Dinoire. A call went out that the team was searching for a donor. When a match was found, Ms Dinoire was operated upon. The initial response, here from *The Lancet*, was positive:

> [T]he time comes when enough is known to do an experiment and when an experiment may be the only way to answer the questions that remain. According to press reports, the French surgical team took care to select a patient who was unlikely to benefit substantially from standard restorative surgery and prepared her psychologically for the operation and its aftermath. They proceeded with the approval of French medical and ethical authorities. Although the decision to operate will no doubt be debated, the French team has taken a cautious and justifiable first step.[15]

Debated the surgery was – even though it might be regarded, – in the strictest sense of the term, as a success.

Much of the debate centred on patient choice. Isabelle Dinoire, it turned out, wasn't such a good patient candidate. Soon after the operation reports started flowing that Ms Dinoire had overdosed on pills following an argument with one of her daughters. Her facial injury occurred as she lay unconscious – her Labrador-cross dog, Tania, gnawing at her mouth, chin, and nose. Ms Dinoire's suicide attempt raised questions about her psychological state, or better put, allowed questions to be raised regarding her psychological state. That the French team operated upon Isabelle Dinoire even though it was possible to raise questions

DOI: 10.1057/9781137452726.0003

about her psychological health suggested for some that they had been working with too much haste, that they were perhaps motivated more by the desire to perform the operation than the desire to treat the patient, and that they cut institutional and ethical corners in the process. The criticism faced by the French team was well known to my interlocutors when they began their work on the face transplant protocol at the centre of my analysis. Indeed, it was in the back and at times the front of their minds during the protocol's development.

Thus when the team began work on the protocol, defining and finding an 'ideal patient' was central to making face transplantation more than merely a technological possibility. It was necessary for closing down the debate as to whether or not the surgery could be performed. In other words, the particular state of health and suffering of the transplant patient could allow the operation to pass as ethical, or not. In particular, the patient's pathology would be so that traditional reconstructive methods were deemed inadequate when it came to improving their quality of life – to adequately restoring their physical, psychological, and social state; but they must also be also physically and psychologically healthy enough in order to minimize the risks of the surgery, especially the risks of biological and psychological graft rejection. Moreover, what was at stake in the domain *writ large* was the very meaning of health and life.

Many anthropologists examining questions of biotechnology have come to draw on the work of Italian philosopher Giorgio Agamben who reminds us that there was no single word for describing 'life' in the Greek philosophical tradition. Rather, they employed two semantically different terms: *bios* referring to an appropriate form of life for groups and individuals that is inherently social, and *zoë* to 'bare life,' the simple act of being alive. This distinction has proved useful for anthropological reflection, driving scholars to examine the ways in which contemporary conceptions of 'life' are continually problematized by scientific and technological developments; how the vocabularies we have are constantly outgrown and made inadequate for discussing emergent forms of life.[16] Face transplant surgery has brought *bios* and *zoë* into a particular relationship.

Those allowing the operation to take place – the patient, the doctors, the State – must be willing to sacrifice the 'bare life' of the patient, to limit the possibility of such sacrifice, and to make it a legitimate one. The notion of quality of life, and the role of bioethics are central here, because what is stake is not only the bare life of the patient, but also

DOI: 10.1057/9781137452726.0003

what an appropriate form of life is. Bioethics emerges as more than a technology for constituting and reproducing the legitimacy of the medical enterprise. It is an especially prominent actor in the production and legitimation of those numerous social sacrifices and medical commitments that must be made in order for the 'good life' to be obtained; meanwhile, understandings of just what is an appropriate form of life are changed in this process.

Bioethical reason and the emergence of 'Quality of Life'

Over the last four decades the 'bioethical enterprise' has extended its reach and influence in the sphere of human subjects research. Through the institutional bodies that it establishes and the guidelines that it espouses, bioethics and (medical ethics more generally) reproduces its autonomy from biomedical practice. Though, as a number of scholars have noted, bioethics is often involved and invoked in such a fashion that it justifies or obscures what would otherwise be seen as unethical conduct. Due to the indeterminacy surrounding just what 'bioethics' is and because of the various 'origin myths' that have resulted from the historical reflections of its practitioners, it is difficult to provide a universal story of the history of bioethics.

When examining bioethics as related to human subjects research, however, its development can be couched in relation to two significant threads of historical experience. The first relates to the role of the Hippocratic Oath as a foundation for medical practice, a text attributed to the Greek physician Hippocrates that stipulates that the interests of patients must always come before those of their physicians, and that doctors ultimately must 'do no harm.' The second relates strongly to the atrocities of Nazi medicine during the Second World War and incorporates a number of ethical codes surrounding medical experimentation that were developed and internationalized for and following the prosecution of Nazi doctors.

The majority of these codes focus on the doctor-patient relationship and the circumstances in which medical experimentation can be carried out. The Nuremburg Code was developed by a U.S. physician advising the prosecution of Nazi doctors and included a set number of principles stipulating the criteria that must be met when performing

DOI: 10.1057/9781137452726.0003

medical research on humans. Updated versions of the Nuremburg Code were adapted and adopted by the World Medical Association, initially as the Declaration of Geneva (1948) and latter as the Declaration of Helsinki (1964). As embodied in these declarations, the focus of bioethics is primarily on the patient-physician relationship and emphasizes the importance of the voluntary, freely given consent of human research subjects; as well as that the potential benefits of all biomedical research must outweigh the possible risks to involved patients. The basis of all attempts to satisfy these conditions was the notion of *Quality of Life.*

An *a priori* general acceptance of quality of life as a measure of health has been used to provide the ethical basis for a potentially fatal procedure that treats people with non-life threatening conditions. The recent emergence of Quality of Life (QOL) as an institutionally situated concept coincides with a movement towards broader conceptions of 'health,' in which the term has come to reference more than the inexistence of illness. Among other things, Quality of Life is a concept embodied in scales that aim to transform the subjective dimensions of health into quantifiable and institutionally legible measurements. Evaluating psychosocial wellbeing is at the core of health-related Quality of Life questionnaires. Moreover, the concept has gained traction as a social and political goal that informs the work of policy makers and expert ethicists. Paralleling this move, medical practitioners have become charged with the task of doing more than simply curing illness; they are increasingly responsible for addressing patient happiness. While restoring physical function is a desired outcome of face transplantation, so is improving appearance and with it happiness and social participation.

The notion of Quality of Life and the various techniques used to assess and define it emerged from early 20th century attempts to measure the very vitality, value, and personality of human populations. In Victorian England, social reformers charged themselves with the goal of measuring and defining human nature. While many disciplinary practitioners forge links between the emergence of Quality of Life and psychometrics, its roots are traceable to the eugenics movement and its view of human nature in which 'the person is largely the sum of a number of potentially measurable abilities and personality traits.' Prominent players in this movement argued that intelligence, for example, was an inherited trait and with it that certain *classes* of people should be encouraged to reproduce at the expense of others, such as the poor, the insane, the criminal. These views remove the impact of social and environmental

DOI: 10.1057/9781137452726.0003

factors, transforming the individual into the sole unit responsible for their success and survival. While the legitimacy and respect of the eugenics movement diminished in the first half of the 20th century, its legacy continues in the design and implementation of techniques used to measure otherwise invisible elements of the contemporary human subject: psychological evaluation.

At the hospital where I conducted the majority of my research on reconstructive surgery in Mexico, many patients and their families are sent to meetings with mental health specialists if a surgeon, a paediatrician, a dentist, or a social worker has concerns about their psychological well being. As part of patient assessment, the mental health worker makes a clinical history. This usually involves carrying out psychometric examinations to test the intelligence, aptitudes, personality and attitudes of the patient. The development of such tests at the beginning of the 20th century correlated both with the emergence of psychology as a scientific discipline and its shift in emphasis from examining the work of the human mind to an examination of difference between individuals. Whether they are 'objective' tests with closed questioning or 'projective' tests with their 'free response measures,' the evaluations that are used by the staff at the División de Psiquiatría y Salud Mental have numerical scoring systems and the results are assessed in relation to standardized scales.

Central to the work of mental health experts in the hospital, the tests provide a way of producing truths about people by allowing the otherwise unknowable and subjective world of the individual to be rendered visible and calculable. Critical analysts of psychology and psychiatry have argued that the test as a standardized form of measurement is also a way of making people classifiable. The test in itself is not able to reveal truths about the patient. Rather, it makes the subjective knowable by allowing patients to be incorporated and analysed within a broader hierarchical and normative system of classification. Tests allow for risks to be measured and pathology to be marked as the scores of patients are situated in relation to a scale of normal and abnormal. Moreover, they allow for symptoms to be seen and for syndromes to be searched for. That is, they provide a method of diagnosing mental health conditions.

Mental health conditions are also classified, and there are standardized ways of searching for them. While this does not mean that diagnosis of psychological or psychiatric disorders are necessarily a smooth process, such testing and such classification of pathology formed part

DOI: 10.1057/9781137452726.0003

of the care offered by the División of Psiquiatría y Salud Mental in the hospital where I spent many hours investigating the everyday workings of reconstructive surgery. Testing provides a means for deciding how to best support patients and their families. This is important because, as the head of Psychiatry at the hospital reminded me, becoming the subject of reconstructive surgery introduces into the lives of patients a completely different type of abnormality, one that confounds their pre-existing sense of difference, shame, and suffering.

In her ethnography of Quality of Life measures in relation to Dystonia, Laura Camfield (2002) argues that these scales work to place the burden of responsibility onto disabled individuals. Like psychometric examinations, the measures seek to make visible otherwise unknown truths about the lives of individuals. By evaluating biological and psychological health, they locate and make open to analysis the very vitality of persons. In doing so they discount the temporal and fluid environmental factors that influence peoples' lives. As Camfield writes:

> If the main determinants of QOL are social and environmental... it can be improved by action at the level of society, but this is obscured when it is located in the 'natural' body of the individual... This reinforces the 'decomposability' of society by reducing 'social processes and structures' to individual attributes and 'detach[ing] knowledge, action and events from their social settings.'[17]

Quality of Life is on one hand an 'explanatory science,' a technology of political legibility that makes readily available for analysis a range of phenomena through the use of specific explanatory principles. The result is a thinning out of complexity resulting in inaccurate representations of lived experience.

Such thinning out of the world in order to make it easily visible is central to contemporary forms of governance. The rise of 'audit cultures' around the world as the primary way in which individuals and institutions are held to account has led scholars to examine the standardization of measurement scales as a principle instrument in 'ethical governance.' In her analysis, Camfield situates Quality of Life squarely within this domain suggesting that it is often as much about efficiency and cost as it is about care. Though the ethical flavour of Quality of Life also exists in its utility to show the efficacy of medical procedures, which has turned it into a resource for championing expensive treatments and rendered clinicians as advocates. This was seen in following the advent

DOI: 10.1057/9781137452726.0003

of cyclosporine as surgeons used the measure to compare life after organ transplant to life on dialysis:

> The objective [was] not merely to facilitate clinical decision making, but also to influence public policy as it applies to federal insurance coverage for kidney transplantation. Favourable estimates of the QOL of transplant recipients augment studies of comparative treatment costs, which show that the initial expenses associated with transplantation surgery are recovered quickly in comparison with the ongoing costs of dialysis.[18]

Quality of Life measures thus embody and reproduce a particular ethic of *responsibility*, in which individuals are held to account *and* incentivized to forge new normative terrains within existing socio-political contexts. More importantly here, they embody and reproduce the very understandings of the individual self upon which plastic surgery *writ large* has gained its legitimacy as a medical practice. The notion and measurement of Quality of Life is arguably more beneficial than not in the context of face transplant surgery, a place where patient death is a reality, because it provides the very ground upon which surgeons have been able to forge their claims.

Notes

1   Morris, P., A. Bradley, L. Doyal, M. Earley, M. Milling, & N. Rumsey (2003). *Facial Transplantation: Working Party Report*. London: The Royal College of Surgeons of England, p. 4.

2   Naugler, D. (2009). 'Crossing the Cosmetic/Reconstructive Divide: The Instructive Situation of Breast Reduction Surgery.' In M. Jones and C. Hayes (eds) *Cosmetic Surgery: A Feminist Primer*, pp. 225–238. Farnham, UK: Ashgate.

3   Gilman, S. (2000). *Making the Body Beautiful: A Cultural History of Aesthetic Surgery*. Princeton, NJ: Princeton University Press.

4   Pigott, R. (1995). 'Aesthetic Reconstructive Surgery.' *British Journal of Plastic Surgery*, 48(5), 338–339.

5   Gilman, S. (1998). *Creating Beauty to Cure the Soul*. Durham: Duke University Press.

6   Hacking, I. (1990). *The Taming of Chance*. Cambridge, UK: Cambridge University Press.

7   Good, M-J. (2001). 'The Biotechnical Embrace.' *Culture, Medicine and Psychiatry*, 25(4), 395–410.

DOI: 10.1057/9781137452726.0003

8   Cohen, L. (2005). 'Operability, Bioavailability and Exception.' In A. Ong and S. Collier (eds) *Global Assemblages: Technology, Politics, and Ethics as Anthropological Problems,* pp. 79–90. Oxford: Wiley-Blackwell.

9   Ibid.

10 Lock, M. (2001). *Twice Dead: Organ Transplants and the Reinvention of Death.* Berkeley: University of California Press.

11 Ad Hoc Committee of the Harvard Medical School to Examine the Definition of Brain Death. (1968). 'A Definition of Irreversible Coma.' *JAMA,* 205(6), 85–88, p. 87.

12 Cohen, L. (2005). 'Operability, Bioavailability and Exception,' p. 85.

13 Infante-Cossio, P., F. Barrera-Pulido, T. Gomez-Cia, D. Sicilia-Castro, A. Garcia-Perla-Garcia, P. Gacto-Sanchez, J. Hernandez-Guisado, A. Lagares-Borrego, R. Narros-Gimenez, & J. Gonzalez-Padilla (2013). 'Facial Transplantation: A Concise Update.' *Medicina Oral, Patologia Oral Y Cirugia Bucal,* 18(2), 37.

14 Slatman, J. & G. Widdershoven (2010). 'Hand Transplants and Bodily Integrity.' *Body and Society,* 16(3), 69–92.

15 The Lancet (2005). 'The First Face Transplant.' *The Lancet,* 366(9502), 1984.

16 Agamben, G. (2000). *Means Without End: Notes on Politics.* Minneapolis: University of Minnesota Press.

17 Camfield, L. (2002). *Measuring Quality of Life in Dystonia: An Ethnography of Contested Representations.* Phd Thesis: Goldsmiths College. p. 218.

18 Joralemon, D. & K. Fujinaga (1997). 'Studying the Quality of Life After Organ Transplantation: Research Problems and Solutions.' *Social Science and Medicine,* 44(9), 1259–1269, p. 1260.

DOI: 10.1057/9781137452726.0003

2
Institutionalized Personhood

Abstract: *Drawing on my research with a group of Mexican reconstructive surgeons, I analyse how the broad debate surrounding face transplant surgery framed the team's relationship with a patient who presented with severe facial disfigurement. According to the team, this patient was an 'ideal candidate' for the operation, a person whose particular state of health and suffering made face transplantation the preferred treatment option. I show that in seeking to treat the patient using the operation, the medical team came to produce a new imperative on the part of the Mexican State to extend its constitutional entitlement to protect the health of its citizens. This extension would allow for donor facial tissue to be harvested, for the operation to be performed, and for patients to be operated upon. In presenting this case, I show that what is at stake at in the realm of face transplantation is more than the health of individual patients; it includes broader issues of how governments should respond to medical advancement.*

Taylor-Alexander, Samuel. *On Face Transplantation: Life and Ethics in Experimental Biomedicine.* Basingstoke: Palgrave Macmillan, 2014. DOI: 10.1057/9781137452726.0004.

Patient selection: An initial attempt at stabilization

The team became responsible for treating Escobar at a time when the assemblage of institutional and biomedical technologies had begun to coproduce what the Mexican team refer to as an 'ideal patient' in the realm of face transplantation. Surgeons, bioethicists, and others working in different medical centres and regulatory bodies around the globe continued to struggle with what *Nature* deemed an 'ethical nightmare.' As already noted, the core ethical issues related to both the effects of immunosuppression and the possibility of rejection, whether psychological or biological. Further, as the Working Party of the Royal College of Surgeons of England emphasized in response to the French surgery:

> Although cosmetic and reconstructive surgery is often reported by recipients to be beneficial in the short term, it is by no means a universal panacea for appearance-related problems. In the absence of studies involving long-term follow up, the jury is still out on whether appearance-enhancing surgery produces lasting psychological benefit or improvements in social functioning.[1]

The working party continued by noting the reported long-term benefits of cognitive behavioural interventions for the lives of people suffering severe facial disfiguration. In contrast to surgical intervention with its aim to cure the psyche by altering the body, such cognitive therapy focuses on changing the very thoughts and focuses of affected patients. Moreover, the report pointed to a catch-22 in the realm of face transplantation: the need 'to proceed with research in order properly to establish the risk of experimental intervention.'[2]

The working party's response was both an overall cautioning of the use of the procedure and the introduction of specific guidelines for selecting 'suitable candidates for face transplantation.' These were organized into six key points: (1) the patient's motivation for seeking the treatment and their expectations of the outcome, which must not be unrealistic; (2) their prior engagement with other, 'less risky' potential solutions, especially psychological intervention; (3) the psychological stability and capacity for readjustment of the prospective patient. That is, they 'should be sufficiently resilient to cope with the considerable stress associated with the transplant, including the "unknowns" associated with a new procedure of this nature, the complex immunological and behavioural post-operative regimen, [and] the risks of rejection…'; (4) their 'level

DOI: 10.1057/9781137452726.0004

of cognitive functioning…should be sufficiently good to understand and assimilate complex risk/benefit information'; (5) the patient must demonstrate 'adherence credentials,' 'as adherence to a complex post-operative regimen is crucial to the successful outcome of the transplant and as some degree of non-adherence occurs in up to 46% of patients undergoing other types of transplant'; and (6) the patient must have a strong social support network in order to 'buffer' them against stresses during and after the procedure.[3]

Ironically, central to the selection of patients is their psychosocial health, the very element of their being that the operation seeks to improve. A suitable patient in this instance is someone who exists in a certain state of psychological health and suffering. They must be physically ill enough to justify the condition but psychologically adequate to take on the responsibilities required for a patient of an experimental biomedical procedure. This responsibility is constituted through relationship with the transplant team, who

> should be satisfied that candidates are capable of assessing whether possible improvements in *quality of life* outweigh the potential morbidity and mortality caused by long-term immunosuppression and the stress of potential or actual rejection of the transplant.[4]

In this instance, the suitable candidate for face transplantation is someone who can and must assume responsibility for deciding and ensuring, to the best of their ability, that the operation will improve their quality of life.

When the team in Mexico became responsible for treating Escobar, they drew on these guidelines, and extended them to include analysis of the physiology of potential patients. The Working Party Report and the case of Isabelle Dinoire were central to how they viewed Escobar; meanwhile the requirements of institutional bodies in Mexico saw them enter into a process of producing their own guidelines for patient selection. Both sets of criteria involve a set of procedures through which transplant teams become responsible for selecting and *shaping* potential patients in order to make them 'suitable' or 'ideal.' It is only through interaction with patients that transplant teams can assess their suitability. This interaction qua assessment is a gauge through which a mutilated person is incorporated into an epistemic system, a place where they are tested as and made into knowledgeable and responsible citizens.

DOI: 10.1057/9781137452726.0004

Operationalizing face transplantation

Dr Haddad was the first surgeon that I spoke to regarding face transplant surgery and one of the two principle surgeons responsible for restoring Escobar's face. When we met, the narrative that he presented to me mixed anxiety with excitement, and revealed a melding of personal and institutionalized ethics. Having already familiarized myself with the debate surrounding the operation, my primary interest at the time was to learn about the motivations of the doctors working on the protocol, and how its development had played out. This was towards the end of 2008. I was told that the development of the Mexican protocol began when the team became, in the words of Dr Jose Haddad Tame, 'serendipitously' responsible for treating Escobar. There were no previous plans to develop a face transplant protocol; it was his condition that was the catalyst for their endeavour: 'we had a patient that was a good candidate probably about a year and a half ago and we worked very hard in order to try to treat this individual [using the operation].'

It was some time after Escobar arrived in the hospital before he was referred to the plastic surgery division to assess his reconstructive needs. In the months prior to their meeting, he was in a precarious condition that required constant care and monitoring. When his health was steady enough for physicians at the hospital to start thinking about facial reconstruction, he was still in a parlous state. As his other treating surgeon, Dr Raymundo Priego Blancas said:

> He had a lot of problems, he almost lost his life during all of this, but well, he is moving forward. He lost almost forty kilos. It was really, really severe. We are talking about a *campesino*, a person with a relatively low cultural background, who doesn't have access to all of the information that we have in cities or in other communities. However, this man, when we saw him, well he had a serious problem in having to restore all of this, because of the exposed jaw bones and the tongue, some of the floor of the mouth was also taken by the lesion. It was practically impossible to do this using a flap or various flaps because it would not adequately restore function. Because of this, we began working with the idea [to do the transplant]. He was a patient that we would consider as an ideal [transplant] case: A young, productive man, with a serious facial condition. So, we spoke to him about the possibility of doing a transplant and, contrary to what we expected, he understood us very well. He understood perfectly what the transplant would involve, that it would take, practically, the face from one person and put in place of

DOI: 10.1057/9781137452726.0004

his. And that this would also involve taking a lot of medication for the rest of his life and a total change to his life, right, to have to change where he lived, worked, etcetera.

Upon viewing and assessing Escobar, the team came to feel obliged to use face transplantation. When talking about the project with Dr Haddad he presented the story in a way that suggested Escobar's condition *obliged* him and his colleagues to develop the protocol. When I asked Dr Priego as to whether or not his condition brought with it such a sense of obligation, he offered me the following response:

> Were we obligated to develop it? Not only for him, but for all of the patients that we have with the same problems as him. He was the catalyst for the protocol because we where unable to find a way to treat him with the options that we had, right? And we considered that the best thing for him was to give him the possibility [of receiving the transplant]. This involved developing the protocol, gaining authorization, and [answering] a series of questions. When we started to develop the protocol we didn't know if we were going to be able to carry it out with Escobar. I mean, with all honesty I'm saying this, we didn't know if we were going to be able to do the transplant with Escobar, but it was worth the effort to try and to have everything ready, if not for him than for another patient; that we would be ready to offer the treatment as the best option. So, yes, we were obliged to develop it at some point [*en algun momento*].

The enfolding of technical, bureaucratic and ethical understandings framed the way the team viewed Escobar and their responsibilities toward him. He was otherwise healthy, relatively young, understood what was involved, and had the support of his family. His face could not be satisfactorily reconstructed without the transplant; he would not be able to adequately eat or speak without it, and he would be subjected to social and psychological suffering. They felt that transplantation was the best way to restore his face. By this time they knew that the operation was medically possible and the specific relationship between Escobar's health and his pathology corresponded 'serendipitously' with emerging criteria for making the operation possible.

They began developing a protocol specifically for him, working long into the evenings, holding meetings with different specialists every few days, and establishing the relationships necessary to make the surgery a possibility within the context of the hospital, and within Mexico more generally. Again, to quote Dr Priego:

DOI: 10.1057/9781137452726.0004

Initially there was a lot of anticipation [*expectativa*] surrounding the possibility of doing it. We worked really hard for a number of months and we were really excited about performing [the operation]. In the end it was a very ambitious project, right, and we believed that we were ready from the technical standpoint, from the legal standpoint, and with regard to all the little details, [such as] medications. I mean … it was an arduous process because we had to search for a provider of the pharmaceuticals, people that would commit to giving this man a life long course of treatment without receiving a penny.

The team was working to gain approval from the ethics, the research, and the transplant committees, which all Mexican public hospitals are required to have in place if medical research and/or organ transplantation is to be carried out within their walls.

Following the successful submission of the protocol to the committees, in collaboration with hospital management, the team applied to Mexico's national sanitary agency, the Federal Commission for the Protection against Sanitary Risk (COFEPRIS), seeking the requisite licenses to perform the operation: licenses to harvest and transplant composite facial tissue. Because the request was the first of its kind in the country, the COFEPRIS opted for an extended review period. The endeavour to treat Escobar with transplanted facial tissue had morphed into a project that would lay the foundations for face transplantation in Mexico.

During this time, it was becoming increasingly clear to the team that their patient was unlikely to receive the care that they wanted to offer him. With each day that Escobar's bones remained exposed, the possibility that he was going to receive the transplant diminished. The doctors were aware from the outset that it was possible that their work would not provide their patient with the opportunity that they hoped, and they were increasingly aware that they were not working simply for Escobar alone but for future patients, too.

In the end, time did run out. When I first met the team members well over a year and a half after the process started, they still had not received final word from COFEPRIS. Waiting gave way to frustration and the moment came when they were forced to operate on Escobar or risk further complicating his condition. They operated on Escobar using the methods and resources that were available to them. They would have preferred to use the transplant. In my interviews it became clear that the doctors knew that they had done the best considering the circumstances, even though there was some feeling of disappointment – if for no other reason than the outcome of the methods they were forced to use.

DOI: 10.1057/9781137452726.0004

Even though Escobar was no longer the centre of the ongoing effort to implement the face transplant protocol, his initial condition, marginalization and disenfranchisement had become the catalyst for the production of a series of arguments and artefacts that sought to make other, similar patients, operable. One of the first steps for doing this was delineating and stabilizing just who, exactly, could and should be operated upon using the operation. Answering these questions entailed working through the bureaucratic guidelines governing clinical research in Mexico, and elsewhere around the world.

On stabilization

The Working Party Report guidelines for patient selection stressed the psychological acumen of potential recipients for face transplant surgery. Both the British and the Mexican cases delineate a particular type of patient-citizen who alongside the transplant team is responsible for ensuring the success of the operation. There are a number of accounts of how in the context of various entanglements, catastrophes, and research terrains, the biological substance of populations comes to mediate their relationship to and/or claims of the nation-State, resulting in citizenship being both performed and reconfigured in new and novel ways. This begs the question, are psychological states a parameter through which citizenship relations are shaped?

Drawing on the English guidelines, the Mexican team outlined six potential patient groups. The delineation of these categories demonstrates that the interface of the psychological and biological understandings that provide the basis of face transplantation as a therapeutic resource may be central to other contemporary manifestations of citizenship. The first two groups presented by the team were possible recipient patients. These two groups consist of acute patients with traumatic facial disfiguration that originated from accidents, necessary surgery, or criminal activity. By selecting these as potential candidates, that team sought to minimize the potential risks and at the same time increase the benefits of the transplant. By selecting acute patients who have not received any prior facial reconstruction to treat their injury, the risks of the surgery are reduced. For example, two of the other patient categories include patients whose disfigurement was caused by the same means as the two operable groups but that have already begun treatment using classical

DOI: 10.1057/9781137452726.0004

reconstructive method. These patients will not be operated upon because they have more to lose – that which they already have gained through their reconstruction – if the transplant was to fail.

The remaining groups categorize patients with prior medical conditions – including psychological instability – that make them too risky to be operated upon using face transplantation, or patients that want the surgery for no other reason but to modify their appearance. The protocol states that the team can see no justification in performing the operation on the latter patients. This categorization of patients only provides the outline for selection of potential recipients. There are other questions to be taken into account. The additional selection criteria seek to further minimize the risks of the surgery. This method of reconstruction will not be performed on all patients from groups one and two, but only patients in a certain state of health and suffering.

The criteria for selection range from 'having a condition that can be effectively treated using the transplant' and being 'compatible with the donor from whom the organ or tissue will be taken' through to demonstrating 'the capacity to tolerate the transplant through the various stages of its development.' This capacity is to be measured through physical, social, psychological, and psychiatric assessment. There are a number of criteria for excluding patients from the protocol, including physical symptoms or forms of illness that could interfere with the medical success of the transplant, to the patient's social life, dependencies on alcohol, pharmaceuticals, and tobacco, inadequate living conditions or familial support, previous attempts at suicide, any demonstrated severe psychiatric or psychological condition, and, of course, ability to give informed consent. What emerges from these criteria is a picture of an *operable*[5] and ethical patient; a person whose certain state of suffering allows the procedure to take place considering the relevant bioethical and technological constraints.

The categorization and exclusion of potentially risky subjects, then, is central for allowing the operation, and the protocol, to pass as ethical. During my fourteen-months of ethnographic research in Mexico, I spent time interviewing a number of bioethics and research officers. Viewing this material alongside policy documents and the publications of the country's National Bioethics Commission revealed a connection between bioethics and Mexico's ongoing aspirations for development. On a national level bioethics is imagined as a mechanism for ordering science and society to help move Mexico into development. Within this

DOI: 10.1057/9781137452726.0004

imaginary and the practices that it engenders, bioethics codifies and seeks to discipline science and surgeons, medicine and patients. In doing so it turns the clinic into a particular rendering of the nation with its own democratic, political subjects. Moreover, bioethics is seen as a way to modify the very fabric of human relations by ordering connections between individuals, institutions, and nations.

As a technology for ordering society, bioethics plays a significant role in the institutionalized imagination of bioethical patient subjects. As many scholars have noted, bioethics guidelines and principles are based on a particular understanding of the human person – the democratic and autonomous western individual. The guidelines for patient selection reflect this conceptualization and demonstrate the explicit relation between bioethics and the State. The categorization of patients and the selection criteria act as a way to minimize the risks and increase the benefits of the protocol. These are risks not only to the patient-recipient but also to the surgeons, science, and the Mexican State. The biological and psychological health of the patient comes to be central to reducing the risk of graft rejection and to increasing the likelihood that the transplant will be accepted by the patient; this is measured not only in terms of a healthy graft but also through the patient's adjustment to having a new appearance and to having new (i.e. someone else's) flesh. The risks and benefits of the surgery are respectively minimized and maximized through the *fresh* biological damage to the patient's body and the potential damage this could cause to their psyche.

The categorization of potential face transplant recipients and the further criteria through which such patients should be selected was one of a series of movements though which the team sought to make the surgery a credible treatment method. This particular movement demonstrates the ways in which bioethics' risk and benefit analysis emerges as a standardized method of calculating and assessing the legitimacy of scientific and medical endeavours, and the large role that it plays in mediating access to emergent treatment methods (and to peoples' bodies) in the process. One of the ways that bioethics works is by helping to establish and legitimize a particular way of seeing illness and attempt to treat it. Such vision comes to be shared both by practitioners and those responsible for assessing their claims.

Bioethics and the practices it engenders come to mediate access to bodies as a ground of knowledge production. Whether through risk-benefit analysis or the need to produce scientifically credible research,

DOI: 10.1057/9781137452726.0004

it plays a significant role in delineating certain patients as more suitable than others. In charting the ethical variability inherent to global clinical trials, for example, Adriana Petryna demonstrates how untouched, 'treatment naïve' populations have been afforded a new value.[6] With bodies untainted by previous pharmaceutical treatment, they are attractive research participants because they are less likely to impact the outcomes of clinical trial regimes. With fresh biological damage and their borderline psychological state, the ideal patients of face transplant surgery are loaded with similar experimental value – in all of its political and medical relevance.

Bioethical citizens

Historians of organ transplantation locate the beginnings of the practice in various historical times and figures. Whether tracing its origins to the work of Aristotle or the enquiries of 13th century Italian surgeons, authors often offer the notion of vivisection to conceptualize the emergence of the practice. From this vantage the history of face transplantation can be located in a broader trajectory of western science and medicine that is intimately tied to politics. The story of transplantation is a story of science. And, as anthropologist Shiv Visvanathan writes, 'The experimental method so crucial to modern science is not only a system of political controls but it incorporates a unique notion of violence – that of vivisection. Within such a framework, the laboratory becomes a political structure and the basis of a wider vision of society.'[7]

Patient selection in face transplant surgery is inherently political. In Mexico, federal policy provided both a ground upon which to appeal for State support for the procedure and the bureaucratic structure through which this appeal was to be made. The arguments put forward by the team needed to extend beyond the medical viability of face transplant surgery as a method of treating patients suffering from severe facial disfigurement. Face transplantation had to be a politically viable treatment option, too. The development of the protocol by the team brought together a series of legal, medical, and bioethical criteria and findings into a novel and productive relationship in which the obligation of the Mexican State to respond to medico-scientific developments when caring for its citizens was called into play.

DOI: 10.1057/9781137452726.0004

The primary basis for doing this was Article 4 of the Mexican Constitution, which offers 'the right to the protection of health.' The very significance and meaning of this constitutional guarantee was effectively recast as the team continued to formulate a new argument about what health is and thus how the State ought to protect its citizens. The team knew this. Or, at least Octavio Madrid Mata, the team's lawyer, did. Sitting in the living room of his home, he explained to me that 'with each moment in which rights theory is advanced, developments in our research continue, and new medical findings emerge, *we have to ask for new rights*' (emphasis added).

In asking for new rights, the team came to engage with the 'interpretative flexibility' of Article 4 of the Mexican constitution. In contrast to the Mexican State's active attempts to limit this flexibility and with it its constitutional obligations, the lawyer and the team more generally began actively engaging with it in an attempt to expand the entitlements offered under the provision. These new entitlements would include the right to harvest donor facial tissue, to perform the transplant operation, and to be operated upon.

The constitutional provision was thus presented as the basis for the protocol as both a form of medical treatment and a method of clinical investigation:

> It is important to note (that whatever the origin of the problem) the basis for any act of clinical research is article 4 of the Constitution, which guarantees the right to the protection of health; that is to say, the development of and research into clinical procedures *ad hoc* that seek to remedy a specific problem, one which falls within the scope of general health in the Republic, is, broadly speak, part of the fulfilment of this guarantee:
>
> We must keep in mind that, from the medical standpoint, the importance of the proposed study and of the transplantation of facial tissue more broadly is its nature as a form of reconstructive surgery that, we insist, does not only treat facial disfigurement but in doing so seeks to restore the anatomical *integrity* of the patient, thus improving their quality of life.
>
> We cannot forget, moreover, that the patients at the centre of the protocol form a specific vulnerable group and, in terms of the constitutional protection of health and of the legal norms at play, should be the beneficiaries of specific health programs.[8]

The duty of the State to permit the protocol was coupled with arguments regarding its ability to do so. Octavio Madrid Mata presented national health law and international bioethics guidelines as the pillars of the

DOI: 10.1057/9781137452726.0004

State's moral authority under which the protocol could be granted its necessary provisions. He emphasized the ways in which the proposed research conformed to the criteria under which human subjects research can be deemed ethical. Moreover, I argue, establishing the scientific and ethical integrity of the protocol was central to its progress; even if the central aim of bioethics and the bureaucracy in which it is embodied and embedded is to protect the integrity of human subjects, it also acts as a tool for maintaining the integrity of science and the State.

The process of establishing the authority and duty of the State to permit the protocol reproduced within it a particular kind of democratic research subject. The integrity of the proposed research and of the State under whose authority it would be performed was dependent on the physical, psychological, and political integrity of the protocol's research subjects – integrity that was to be both maintained and restored. Patient selection was of course a central factor here (the rationalities of which I explored above); but so was informed consent. Though, there was a need to establish just what informed consent meant in the case of face transplant surgery and how it could be gained.

The arguments that were put forward in the protocol suggested that informed consent in this case should involve the production, transmission, and acceptance of knowledge and responsibility in and between the patient, the doctors, the State and those supplementing it. The protocol put forward a number of criteria suggesting just what informed consent should consist of in the case of face transplant surgery, and that there was a need to establish a specific institutional context in which these criteria could be evaluated in order to allow such consent to be gained. In particular, the ability of the patient to act as an autonomous subject would have to be established and their action as such a subject assessed: 'in order to avoid any dependence on the transplant personal the ad hoc formulation of a separate committee that will oversee the medical attention, seek informed consent, and provide information regarding the psychological state of the transplant candidates has been planned' (Madrid Mata 2007: 39).

Informed consent as it exists in the protocol is both a means to and a mode of citizenship. It implies a series of responsibilities on the part of both the Mexican State and the patient in and around an ongoing medical commitment. The State is required to provide the information, environment, and resources (such as pharmaceuticals, even if they are procured through a public-private partnership) necessary for the act of

DOI: 10.1057/9781137452726.0004

informed consent to be a valid one. Along with meeting a set of criteria through which they are deemed able to provide informed consent, the patient is also imbued with a set of responsibilities: they must be a knowing subject that understands the procedure and its consequences such as the risks of immunosuppressive therapy, the possibility of irreversible rejection, and any psychological difficulties that could emerge; they must accept (and manage) these risks; and, they must promise to continue a life full of social, biological, and familial sacrifice (c.f. above comment of Dr Priego Blancas). Understood in this way, informed consent obliges both the physicians and the patients to act in accordance with a set of procedures based on established knowledge and ways of knowing in order to maintain both their individual integrity and the integrity of those that the informed consent places them in relation with.

The protocol thus produces a new account of the form of life that should be offered to certain reconstructive surgery patients. This form of life is as much political as it is social and biological. The institutional authority of bioethics renders legitimate more than a specific mode of rational decision-making; it provides an authority under which bare life itself can be sacrificed in pursuit of a particular, emerging form of *bios*. In doing so it establishes a number of parameters through which *integral* relationships are established and designed to limit the possibility of death. Moreover, it is central to the production and of patients as bioethical citizens.

The first time I interviewed Dr Haddad in 2008 he mentioned to me that he had at times felt anxious and unsettled when he thought about performing the operation, especially on Escobar. There was still some doubt in his mind as to whether it was the right thing to do. He associated this anxiety with the unknowns surrounding the operation, which he suggested were responsible for the slow global uptake of the operation as a treatment method. And he further associated the first operation, the French case of Isabaelle Dinoire, with the egocentrism of the doctors who performed it.

> Now, if you have the possibility to perform an innovative procedure that could be revolutionary and that is going to gain a lot of attention from all over the world, then ... you can skip [*brincar*] the ethical issues in an act of egocentrism, but this isn't right, no? ... This is why the operation hasn't been performed in England, it hasn't been performed in the United States, it hasn't been performed in Mexico.

DOI: 10.1057/9781137452726.0004

Meanwhile, he said, 'the French with the eagerness to perform the first and only [face transplant operation] skipped many of the necessary steps, and the first and most important is to have selected an ideal candidate.'

Face transplant surgery produces its patients as specific types of research subjects. They become coded as 'ethical' and 'national' research subjects in that they allow science and medicine to progress within the borders of nations, while maintaining the integrity of the State and its doctors. But the term 'research subject' must be used in this instance to think about the subjectivities – the lived experiences, personalities, and emotions – of potential transplant recipients. The subjectivities of the patients, along with their biology, must exhibit certain qualities, must exist in a certain balance between normal and pathological: it is in this balance that they become operable and ethical because they conform to the very guidelines and arguments that (re)produce them in light of current technological possibilities and limitations. And, of course, it is the very subjectivities of these patients that are at play in this new biotechnical domain. Face transplant surgery seeks to restore their biological, social, and psychological state – to make them more 'normal.'

In attempting to make its patients more normal, the field of face transplantation constructs new hybrid human kinds. Recipients and donors emerge, to borrow from Marilyn Strathern,[9] as *dividual* beings – in the latter case persons live temporarily only through the help of machines. In the next section we turn to the lived experience of transplant recipients, to the experimental qualities of this experience, and to how we may incorporate it into an assessment of the field.

Notes

1   Morris, P., A. Bradley, L. Doyal, M. Earley, P. Hagan, M. Milling, & N. Rumsey (2006). *Facial Transplantation: Working Party Report* (2nd edition). London: The Royal College of Surgeons of England, p. 16.

2   Ibid.

3   Ibid.

4   Ibid., p. 18 (My Emphasis).

5   Cohen, L. (2005). 'Operability, Bioavailability and Exception.'

6   Petryna, A. (2009). *When Experiments Travel: Clinical Trials and the Global Search for Human Subjects.* Princeton: Princeton University Press.

7   Visvanathan, S. (1997). *A Carnival for Science: Essays on Science, Technology, and Development.* Delhi: Oxford University Press, p. 22.

DOI: 10.1057/9781137452726.0004

8 Madrid Mata, O. (2007). 'Trasplante De Tejido Facial. Aspectos Jurídicos.' *Cirugia Plastica*, 17(1), 31–48, p. 35.

9 Strathern, M. (1988). *The Gender of the Gift: Problems with Women and Problems with Society in Melanesia.* Berkeley: University of California Press.

DOI: 10.1057/9781137452726.0004

3
Self-formation and Ethical Being

Abstract: *Analysing published interviews with face transplant patients and their families, I examine the profound shifts in individual identity and social relationships that result from the operations performance. In doing so, I demonstrate that face transplant surgery is transformative in a way that otherwise escapes the attention of regulatory bodies. Their primary concern has largely been with issues of psychological rejection on the part of the recipient, and how to retain the anatomical integrity of the brain dead donor's body considering the overt damage caused by removing facial tissue. Little has been said about the psychosocial consequences, good or bad, of seeing the face of your mother, father, child or sibling on another person. I raise the question of whether established modes of evaluating clinical medicine are able to capture the newfound complexity that the operation introduces into the lives of patients, their families, and the relatives of the deceased donors. Drawing on social science notions of ethics and patient knowledge, I sketch a blueprint for incorporating patient experience and decision-making into existing modes of post-transplant assessment.*

Taylor-Alexander, Samuel. *On Face Transplantation: Life and Ethics in Experimental Biomedicine.* Basingstoke: Palgrave Macmillan, 2014. DOI: 10.1057/9781137452726.0005.

DOI: 10.1057/9781137452726.0005

What of the (brain) dead?

> Our primary purpose is to define irreversible coma as a new criterion for death. There are two reasons why there is need for a definition: (1) Improvements in resuscitative and supportive measures have led to ad hoc increased efforts to save those who are desperately injured. Sometimes these efforts have only partial success so that the result is an individual whose heart continues to beat but whose brain is irreversibly damaged. The burden is great on patients who suffer permanent loss of intellect, on their families, on the hospitals, and on those in need of hospital beds already occupied by these comatose patients. (2) Obsolete criteria for the definition of death can lead to controversy in obtaining organs for transplantation.[1]

The first mention of brain death in Mexican legislation occurred in 1973 and by 1976 there were official guidelines for harvesting organs from brain dead people published as part of the General Health Law. While there is no detailed account as to why it took over seven years for the Harvard criteria for brain death to be entrenched in Mexican law, unofficial accounts suggest an incompatibility with local Catholic beliefs. The delayed adoption of the redefinition of death by the Harvard committee was not limited to Mexico. A general discordancy between State-fostered belief systems and the notion of brain death has been paralleled by physician-initiated actions to counter its homogenizing effects ever since its emergence. As well as shifting the meaning of death away from the stopped heart to focus attention on the non-functioning brain, the criteria removed the autonomy of medical professionals to decide on a case-by-case basis whether or not someone was indeed 'dead.' It did so by shifting such decision-making from the realm of the physician and into that of a committee constituted by a number of already acquainted doctors, as well as a lawyer, historian, and theologian. Moreover, 'biomedical research interests were perhaps better served than challenged by the Committee because it institutionalized the involvement of transplant interests in further discussions surrounding "death".'[2]

There are a number of reasons to believe the emergence of 'brain dead' persons was tied to the development of and interests of transplant surgery. While the Harvard Committee emphasized the social and economic burdens of brain damage resulting from advancements in life prolonging activities, a body of background notes and memos suggests its members were concerned primarily with issues of organ procurement. It was in no way the first discussion of how to ethically, medically, and legally harvest

DOI: 10.1057/9781137452726.0005

tissue from patients; indeed the report brought together a number of debates and concerns that had been present since the early 1960s. The criteria that they outlined sought to make available what had previously been called 'live cadavers' for organ transplantation, and to keep the 'dead' within the care of medicine, rather than that of the family.

The report describes the characteristics of 'irreversible coma' and how to assess it. Three selection criteria are central to the patient evaluation. Here 'life' is tested to the extreme: (1) Unreceptivity and Unresponsitivity refers to a 'total *un*awareness to externally applied stimuli and inner need and complete unresponsiveness' so that 'Even the most intensely painful stimuli evoke no vocal or other response, not even a groan, withdrawal of a limb, or quickening of respiration; (2) No Movements or Breathing as witnessed by 'observations covering a period of at least one hour by physicians' that demonstrate it is adequate to satisfy 'the criteria of no spontaneous muscular movements or spontaneous respiration' as well as 'the total absence of spontaneous breathing' to be tested by removing for three minutes the patient from a respirator; (3) No Reflexes, as the 'abolition of central nervous system activity is evidenced in part by the absence of elicitable reflexes' so that the person should have a 'fixed and dilated pupil' and 'Ocular movement (to head turning and to irrigation of the ears with ice water) and blinking are absent. There must be no evidence of postural activity (decerebrate or other).' Further, the use of Electroencephalography to measure electrical activity in the brain is suggested as a confirmatory measure.[3]

Although significant in its continued use in defining brain death, the Harvard Committee report did not settle any debates outright nor increase the number of transplant operations being performed. Concerns continued to plague the 'brave new world' of transplant medicine, as it was dubbed, with public and professional concerns regarding the use of bodies with a beating heart. At this time, it was heart transplantation that was on the agenda in the world of medicine, but this was a death-plagued realm with many organ recipients failing to survive beyond twelve months post operation. It was not until the improved life years and quality of life offered by cyclosporine that transplant surgery grew. Meanwhile, the issue of familial consent for donation resumed in many countries around the world, although some worked with the idea of 'presumed consent' making organs readily accessible for harvest.

Cyclosporine eased criteria for matching organs between recipient and donor by tackling the issue of immune suppression. Another important

DOI: 10.1057/9781137452726.0005

issue in finding a match in transplantation is that the recipient and donor have the same blood type. In face transplant surgery this extends into the very appearance of the human tissue being harvested. Finding a match involves a donor with corresponding, though not the same, facial features, involving things as basic as the skin colour, sex, and age of the brain dead person.

In transplanting facial tissue from one person to another, the appearance of the recipient becomes an amalgamation of the original two faces. For this reason, members of the Working Party of the Royal College of Surgeons wrote:

> Teams should be vigilant for signs of psychological rejection of the donor face, for example, lack of interest in looking at the face in a mirror or indications that the patient feels the new face is 'not the real me.' Interventions to facilitate the integration of the graft into the patient's self-image should be initiated in the early postoperative phase, for example, through self-care activities such as massaging surgical scars. The recipient may need assistance to resolve complex feelings about the donor (for example, curiosity about the sort of person, guilt about the donor's death, gratitude to the family).[4]

The life of the donor continues beyond their brain or total death. Their identity is forged anew as it is moulded onto a new body and enters into a new life. Like other transplant procedures, facial tissue as an organ becomes loaded with significance as it circulates between bodies and is reimagined by family and friends of donor and recipient alike.

In a story for National Public Radio entitled 'Falling in love again,' reporter Melissa Block interviewed face transplant recipient Carmen Tarleton and Marinda Righter, the daughter of her donor. Block begins by enquiring into the 'signing off' process that accompanied the transplantation of the facial tissue. Marinda's mother, Cheryl, was registered as an organ donor. After dying of a sudden stroke she finalised the use of her mother's organs for transplantation:

RIGHTER: I signed off on it when I was at the hospital, and after leaving, I got a phone call from one of my New England donor angels, Dan, and he told me that there is a recipient for my mother's facial tissue. And, you know, at that time, it was actually a no-brainer. I was just kind of blown away by the whole thing. It sounded like something out of a science-fiction novel. The one question that I had for Dan was: Dan, is this person going to look like my mother?

BLOCK: Yeah.

DOI: 10.1057/9781137452726.0005

RIGHTER: And he said: Well, no, because the other – the person has a different facial structure, and that will mould to their face over time. And so that was, you know, that was the right thing to say.

There is a contrast between Marinda's initial concerns and how she now sees the operation. To begin with, it was important that her mother's identity not be transferred, that the donor would not resemble the woman that raised her. As the conversation continued, it became clear that Miranda now searches for aspects of her mother in and through the body of the person that lives with her deceased parent's face. The interview continued as follows, with the host asking about the first time the pair met, 'when you first got to meet Miranda, what did you tell her?'

TARLETON: Well, when she walked into the room, I stood up from my chair, and she came towards me, and she said: Can I hug you? And I said yes. And we just embraced and cried. And it was probably one of the best feelings I've had in my life. [...] She's just such a beautiful person and such a giving person and selfless, as I know her mother was as well.

BLOCK: And, Marinda, what was it like for you to see Carmen having received, you know, the face transplant donated from your mother?

RIGHTER: It was truly a cathartic experience. I saw her through the window at the library at Brigham and Women's, and just seeing her profile, I mean, it hit me. I didn't go in with any expectations. I'm like, I know she has my mother's face, and I'm really excited to meet her. ... And I stared at her, and I think my next question was: Can I touch your face? ... And instant - just an incredible instant connection. Man, I fell in love with Carmen there, and I've never felt closer to my mom in that moment too, so wonderful.

Marinda seems to present her experience of meeting Carmen, whose face is both her mother's and Carmen's alone, as natural and organic. The show's host continues by encouraging Marinda to speak about the aspect of seeing and touching her mother / Carmen after the transplant: 'you feel in some level like you are looking at your mother or touching your mother when you did touch her face?'

RIGHTER: I do. It's kind of a funny story, but my mom, she was kind of obsessed with age and getting older. And she would point out her little age spots, and she'll say, oh, I need to cover up this spot. I'm like, Mom, I never noticed those spots until you mentioned it. And it's funny it was the first thing I noticed on Carmen, and it made me joyous. And Carmen pointed out her little – she had like a little mole on her face, and I said: Yes, I know that. I know that mole. Oh, my goodness, I know that freckle. And even the eyebrows and the nose, you know – though Dan told me she was going to not

DOI: 10.1057/9781137452726.0005

look like my mother, she did, and that brought me a lot of comfort. I miss my mother, but, man, what a gift. And she would have chosen Carmen, and *I think she did.*[5]

It is typical in transplant surgery for donor family members and recipients to evoke the language of the gift of life. A liver, heart or kidney allows the recipient to avoid otherwise certain demise. Meanwhile donor family members often experience the gift of giving.[6] Unique in the above conversation is the social gift afforded to Marinda: to be able to see and experience her mum anew. The operation becomes transformative in a way that otherwise escaped the attention of committees and review boards involved in the regulation of the operation. Their primary concern has largely been with issues of psychological rejection on the part of the recipient, and how to retain the anatomical integrity of the donor's body considering the overt damage caused by removing facial tissue. Little has been said about the psychosocial consequences, good or bad, of seeing the face of your mother, father, child or sibling on another person.

In this context, individuals are not mere bodies and minds but social beings whose biological stuff is made to live on past death. The mind body-duality that underscores medicine and is enmeshed in the very operationalization of face transplant surgery is refigured and emerges as a composite of various minds and various bodies, alive, dead, and in-between. Persons are no longer individuals. They are *divided* up into pieces that are then moved onto the bodies and into the lives of others. Carmen Tarleton's initial reaction to the possibility of receiving transplanted facial tissue was conflicted and along with the rest of her family she found it 'creepy' and 'weird.' Her daughter Hannah found it difficult to reconcile the idea of her mother looking different, even though she had been left extremely disfigured by a domestic attack involving acid and beating. Carmen said, 'Imagine if you had my face, and you couldn't smile or kiss, and people didn't know when you were crying or laughing. But the biggest thing is blinking eyes. I want to save my sight for as long as possible.'[7]

Face transplant surgery brings together social relationships and experimentation. These relations extend beyond immediate family and friends, or the relatives of donors; they are situated in the context of medicine, law, and policy. The story of Carmen Tarleton reminds us of this and in doing so the limitations of formal reasoning in biomedical practice. The autonomous individual is at the centre of bioethical

DOI: 10.1057/9781137452726.0005

thought. The stories of Escobar in Mexico and Carmen in the USA alert us to the inherently social nature of face transplantation, and experimental biomedicine in general. While patients are shaped as responsible citizens with almost contractual obligations, their family members are enrolled as dependent carers, as persons who must to the best of their ability accept and adapt to the choices of their loved ones with all their experimental value.

Joy and disgust and being a hybrid

The first hand transplant patient could not live with his hybrid state. He was unable to make the hands of a dead person his own, and so he was left with no hands at all. To whom does Carmen Tarleton's new face belong? Answering this question takes us back to the very first and most controversial face transplant surgery. When being interviewed for a book on her operation, Isabelle Dinoire recounted the disgust that she often feels when reflecting on living with a suicide victim's face in place of her own. Especially troubling is touching the inside of the other person's mouth whenever she raises her tongue, and seeing the hairs that sprout on her chin, something that did not happen before the surgery: 'You know it's yours but at the same time "she" is there ... I am making her live but that hair is hers.'[8]

That Ms Dinoire has experienced identity trouble post-surgery is perhaps something of an understatement. Her face is a composite, a hybrid in which the dead woman is able to live on both through the biology of the transplanted flesh and in the very thoughts, feelings and personality of her host. Writing from the field of science and society studies has demonstrated the importance of locating non-human things within the web or network of social relations in which people live. When looked at in this way, non-human 'actants' influence the functioning of society and what it is to be a person today.[9] The above characterization seems to afford a sort of parasitic agency to the tissue of the dead donor. As skin, hair, and mouth lining are able to live on, away from the being in which they were originally a part, they exist as something both and not fully 'human.' Through ongoing medico-political acts they are biologically and socially fused to the organism to which they are transplanted, to a body, which is unable to reject them because of its drug-suppressed immune system.

DOI: 10.1057/9781137452726.0005

It is not the recipient's disgust but the donor family's joy that is at the fore of Carmen Tarleton's story – at least in the above extract. If her comments suggest an overall acceptance and ability to adapt to her new look then they also demonstrate a multitude of negotiations taking place in her post-transplant life. In this life she is part of a new network of social relationships in which different people claim a stake. Numerous bioethicists have suggested that more so that with other organ transplant procedures, face transplant donors bear a level of responsibility towards the donor's family. Among the reasons for this is that the procedure deprives grieving families of the ability to say final goodbyes in open casket ceremonies. Patients like Carmen Tarleton should do their upmost to ensure that such sacrifice is not in vain. A different picture of responsibility exists in the above interchange between her and Cheryl Righter's daughter.

The joy of Marinda Righter is accompanied by a plethora of speech acts that see her mother live on both vicariously and literally on the face of another person. The parasitic quality of the donated tissue is more outright social here: while the same biological features are similar to the first case, the words and joy of Marinda give life to her mother as she sees her freckle, mole, eyebrows and even nose in/on their new host. Her mother is brought back to life vicariously as she *chooses* and then clings to her recipient host. Responsibility here entails allowing someone else to live on through you, by caring for his or her face that is now (sort of) your face. It exists because we humans have been able to conquer the nature within us, to curtail cellular interaction, and thus bring into a transformative relation the biological and the social.

Euro-American kinship itself can be thought of as something of a biological and social composite. It is an institutionalized assemblage of social relationships between people with shared, traceable genetic lineages. New technologies have recast the importance of our genes for understanding who we are by producing new ways to make and conceive of kin relations. When people are confronted with genetic knowledge – that they are carriers of certain conditions, for example – they often don't know what to do with it. Underlying the difficulties of how to react to genetic knowledge is the idea that genes make us who we are. Faces too make us who we are. Transplantation moves genes and faces and in doing so pushes the boundaries of kinship. I am not saying that Carmen Tarleton is taking on the character of mother to Marinda Righter but that the latter's projections, and the movement of

DOI: 10.1057/9781137452726.0005

her kin's biology (including genes) onto another person demonstrates a reconfiguring of relatedness and production of hybrid selves.

Accounting for the future

Institutions today, especially fiscal institutions, are constantly engaged in strategies of managing our expectations. In his position paper in support of face transplant surgery for the blind, Dr Bohdan Pomahač of Brigham and Women's hospital writes, 'With the passage of time, increasing experience with facial transplantation has demonstrated it to be both safer and more effective than initially anticipated. In accordance, we have witnessed an ongoing evolution of what features characterize the ideal facial transplant recipient.'[10] Two of the seventeen patients operated on at the time this was written had died. Along the way, many other patients almost experienced biological rejection that would have required a complete restoration of the face using autologous tissue. The once anticipated future was very precarious indeed.

The previously death-filled hypothetical future has been silenced as it enters the present as a reality, with previously unimaginable forms of practice taking hold and producing a yet to be stabilized ethics. Doctors were concerned that the transplant could take the lives of otherwise healthy individuals; it has done so, yet the procedure continues. The deaths are made to not count against the overall safety of the operation because one happened 'under uncertain circumstances and the other from overwhelming sepsis in the setting of multiple simultaneous allotransplantations.'[11] The first fatality referenced happened in China, and is commonly understood to have resulted from non-compliance with the immunosuppressant regime. The second happened in Spain following a dual hand and face transplant operation. This argument suggests that the ten per cent chance of death is perhaps lower when the procedure is performed in certain circumstances and limited to a single allotransplant.

However, appraisals are being made in the present, where the long-term outcomes of face transplant are unknown, causing some to plead with plastic surgeons to slow down and rigorously assess just what they are doing. A 2011 review of the field, wrote, for example:

> No long-term studies have demonstrated improved quality of life for facial transplant recipients. Nor has it conclusively been shown that current

DOI: 10.1057/9781137452726.0005

immunosuppressive regimens will maintain long-term viability of the transplant. Thus far, evaluation of outcomes is limited to the small number of patients transplanted with the longest transplant only 5 years postoperative.[12]

It is not debated that the operation engenders many kinds of experiment: medical, psychological, social, ethical. The debate is how to make sense of the present and move into a future that may slip away as the value-added-label of experiment no longer applies. The notion of experiment is both enabling and disabling; the value it adds to the operation is both positive and antagonistic. It signals an area of unknowns, a place where knowledge can be produced and potentially capitalized upon in scientific, monetary, and moral terms. As the same review of the field notes:

> The US military has taken a keen interest as well and the recently formed Armed Forces Institute of Regenerative Medicine, dedicated to helping soldiers with burn and blast injuries regrow tissues, has contributed funding to some of the early US facial transplantation efforts. All the press and media coverage surrounding the few cases so far has also led to a relative windfall for some institutions, with investors contributing money and a broad range of public and private organizations donating money.
>
> With the continuing success of facial transplantation, the question will eventually become, will we be *allowed* to perform this surgery. Once the initial fervour over facial transplantation subsides, where will the funding come from? Will insurance companies and third-party payers cover the costs of [face transplantation]?[13]

At the same time, the experimental quality of the procedure moves it into the realm of institutional oversight, as we've seen above. Resulting from the history of medical research around the globe, the individual lives of patient research subjects, quantified through expert measurements and imagined vis-à-vis ingrained understandings of personhood, are guarded through paperwork. Embodied within this paperwork are particular notions of a person.

This person is a psychologically stable self who both looks and feels normal and happy. While both qualitative and quantitative indicators are used in assessing patient suitability, it is the latter modes of measurement that form the basis of postoperative evaluation. In reporting the outcomes of their first face transplant operation, Maria Siemionow and colleagues began by providing an overview of the social and biological life of the patient leading up to the operation:

> [T]he patient had eloped and married young, had worked in a painting and wallpapering business with her husband, and had co-owned a restaurant/bar

DOI: 10.1057/9781137452726.0005

where the assault took place. She was in the painter's union for 7 years. She had numerous hobbies. She was raised as a Christian, but was not active in a church. Mental status examination was not remarkable, as she was alert and oriented, without any hallucinations or delusions.[14]

This social history was used in evaluating the suitability of the patient, Connie Culp, for what was the world's first 'near total' face transplant operation. Qualitative assessment of the patient's life happened concomitantly with quantitative evaluation of her cognitive and psychological suitability. This is typical in the field, as we saw in the Mexican case and the report issued by the Royal College of Surgeons. Yet, the qualitative all but disappears in postoperative evaluation, except when accounting for inauspicious results – note the last sentence below:

> In our case, the patient had a decline in Beck Depression Inventory (BDI) score from 16 to 6 by 3 months after face transplant while still on the [antidepressant] escitalopram. On 12 July 2009, the BDI score was 14 reflecting CMV infection and other issues at home. Appearance self-rating jumped from 3/10 after her injury to 7/10 within 6 weeks of transplantation and is now 7 – 8/10. State-Trait Anxiety Inventory state anxiety score stayed constant as did Rosenberg self-esteem score. Her self-esteem was already well developed prior to the injury and to a large extent not dependent on her appearance, but on pride in activities she engaged in.[15]

The social aspects of patient lives so crucial to the operation of the surgery as a politically legitimate, ethical enterprise, are reduced to numbers. These numbers focus on a specific dimension of the social lives of patients, the psychosocial. Any real sense of messiness, joy, happiness or difficulty, that is the very process of figuring out how to live, is edited away by the systems of making institutionally legible the lives of face transplant patients.

The quantification of the inner aspects of the human selves that are subjected to experimental medicine is part of the production of value in this new realm. Value is a word that has at least two meanings: the economic and the moral. The moral value of medicine exists in the improvement that it makes to human life, improvements yet to be fully quantified over the long-term in face transplantation. It is here that the moral and the economic are brought into a unique relationship, in a manner not dissimilar to the contemporary biotech industry where economic value is produced in conjunction with the promise of improving the health and vitality of populations.[16] Here, though, it is

DOI: 10.1057/9781137452726.0005

the institutionalized manner of calculating Quality Adjusted Life Years (QALYs) that sees the implosion of the monetary and the moral.

A form of measurement developed to aid decision-making surrounding the allocation of scarce economic resources between competing health care projects, QALYs now play an important role in public healthcare systems. In their 2011 review of the 'progress and future targets' of the field, plastic surgeons James Edwards and David Mathews reflect on the operation in terms of QALYs. Drawing on analogous analysis of hand transplant operations, they state that:

> While initial examination of cost-effectiveness may argue for or against facial transplantation, the final determinant of whether this procedure is funded will be how society views the benefits of the procedure. When renal transplantation is examined, the relative monetary cost versus alternative therapy (outpatient dialysis) is initially high but the improvement in quality of life is significant and has been deemed 'worthy' by society. In addition, after approximately 2 to 3 years, the cost of renal transplantation compared with conventional or intensive renal dialysis is much lower, owing to the fact that most of the cost for transplantation is incurred 'up front' […]
>
> For facial transplantation, economic analysis would have to focus on the direct benefits to the patient, such as their improved sense of self and reintegration back into society. The benefits to society have to be taken into account, as patients may be able to rejoin the workforce and participate socially rather than being reclusive and isolating themselves. The first French facial transplant patient, Isabelle Dinoire, has been reported to have gone back to work and is considered to have had successful reintegration into society.[17]

Calculating the value of face transplant in terms of QALYs allows for it to be understood and assessed in terms that make it legible in the context of political decision-making. Here, health is not so much a commodity as a sociopolitical good in that the wellbeing of individuals is worth maintaining especially if it benefits society broadly.

The practice of plastic surgeons seeks to work with and around this institutionalized model of accounting for health. The notions of rights and justice have provided the tools for doing this. In Mexico, the ethical sensibilities of the surgeons responsible for treating Escobar resulted in them entering into a relationship with the State and forging new rights for patient-citizens. Similar, in the USA surgeons have invoked the rights of their patients when laying out the ethical foundations for the operation. Again, Dr Bohdan Pomahač's collaborative position statement on using the procedure on legally blind patients provides a pertinent example of

DOI: 10.1057/9781137452726.0005

this tactic. Moreover, by entering into a conversation about who should be operated on using transplanted facial tissue, the very idea that the surgery should be used as a treatment method is further solidified:

> A related ethical concern is the notion of distributive justice and the alloca-
> tion of scarce resources. The implicit concern in the case of facially disfigured
> blind patients is that it may not be reasonable to expect them to acquire
> benefits of face transplantation equivalent to those of sighted patients or suffi-
> cient to justify the risks of lifelong immunosuppression, and that allocation
> of facial allografts should be based on the notion of providing the greatest
> good for the greatest number. This position is not supported by outcomes
> witnessed to date, nor is it justified by concerns regarding the ability of blind
> patients to comply with required therapy.[18]

Broader political understandings and bureaucratic apparatuses have proved foundational to the development and assessment of the field to date. What has resulted is a limited picture of the complexity of the field, especially as it relates to the lives of its patients.

Who gets to make the ethical?

Anthropologist Michael Jackson has recently written 'In developing an ethics of the intersubjective, we need a method of study that avoids prejudgments as to what is right and wrong, good and bad, and thus draws us deeply into the complexity of everyday situations.'[19] He sketches his stance by drawing on a number of literatures, especially the recent 'ethical turn' in Anthropology, a turn that has seen renewed attention placed on the everyday, minute and ordinary decisions and relationships through which people constitute life worlds. In his writing on the "ethics of the act," for example, Michael Lambek states 'the ethical is intrinsic to human action, to meaning what one says and does and to living accord-ing to the criteria thereby established.'[20]

Face transplant surgeons have been involved in the production of institutional criteria through which the operation has come to pass as ethical in hospital and national bureaucracy. They have entered into conversation in which pre-existing understandings and practices result-ing in the ethics of the operation focusing especially on Quality of Life. I view this action as reflecting an on-the-make form of ethics. As surgeons have continued to perform the operation, so too have the sought to alter

DOI: 10.1057/9781137452726.0005

understandings about the human body itself. In 2011, Maria Siemionow published an autobiographical account of her 'quest to perform America's first full face transplant.' The book provides an overview of personal and professional growth in which understanding the body is intimately tied to and produces the imperative to improve the lives of others. For Dr Siemionow:

> No other anatomical structure compares in complexity to the face. The hand approximates it. Some would argue that the complexity of the hand equals that of the face, but although the hand may have more obvious parts, the face has a more intricate, complicated, and subtle musculature. … Nerves bring impulse to the face muscles, telling them when to pucker, when to spit, and when to squint. These nerves also penetrate the skin to a point near the surface, allowing it to register heat, cold, pressure, and pain.[21]

The face, she argues elsewhere, is an 'organ' – not merely a composite of flesh but a 'differentiated structure comprising tissues that perform a specialized function in an organism.'[22]

Face transplant surgeons have been required to justify their endeavour. Institutional review boards, transplant committees and the like have asked them to stop and think about just what they are doing and their motivations for doing so. The result has been the production of a new way to see the face. Improving appearance is but a small facet of the operation, we're told. Rather, the face is as *vital* as the parts of the body that purify blood or pump it around the body: it is 'an organ.' The operation 'should be considered as an organ transplantation that enhances the quality of life to a degree comparable to that of solid organ transplantations.'[23] This is because:

> One of the fundamental functions of the human face is the ability to receive multimodal sensory information from the environment and to convey it to the cerebral cortex for integration and processing. The presence of normal sensation is important not only for the discrimination of touch, temperature, and pain, but also for initiation of vigilant or defence reactions. The presence of labial sensation helps in avoiding drooling while eating or drinking. Stretching of the perioral skin contributes to the precise articulation in speech. Interestingly, cutaneous stimulation increases the intensity of estimates of the olfactory system. It has also been reported that facial skin cooling decreases the heart rate and increases blood pressure. Finally, normal sensory pathways allow to draw pleasure and satisfaction when exposed to external stimuli. It is clear that restoration of the above functions is expected and essential for the optimal outcomes following face transplantation.[24]

DOI: 10.1057/9781137452726.0005

This unification of a body of composite biomedical literature dealing with the face both reveals an underlying professional ethic and is an ethical act in itself. The importance of the face extends beyond interpersonal relations and how we see ourselves. Siemionow and colleagues present it as integral, vital to our being in the world: alongside sensory capacity and phonetic ability, the face contributes even to cardiovascular wellness. In presenting this argument, the authors are laying out a position that itself calls for action.

As doctors like Siemionow have been working for the opportunity to use the operation on their patients they have been staking a claim about how they believe life should be lived in their local, highly technical, medical worlds. Resulting from their struggles, a new kind of life or death obligation has been cast upon patients. The 'ethics' of the operation is based on reconstructing the mindbody and social life of disfigured patients in accordance with established biomedical understandings about normality. For the operation to be a success, patients are required to act in accordance with the criteria of biomedicine. Siemionow herself has written 'the patient is perhaps more responsible for the success of the procedure than all the medical specialists involved' because the operation requires an ongoing commitment to a healthy lifestyle and immunosuppressant regime.[25] A lot more is required for the operation to be considered a success, as a return to the case of Isabelle Dinoire demonstrates. The patient must comply with a set of criteria established by the international transplant community (and solidified in local medico-bureaucratic contexts) and with a form of self-care that extends beyond eating well and taking your medicine.

The controversy surrounding the first face transplant became a conduit through which surgeons and others engaged in the field were able to stake a claim for themselves as ethical actors. The team of French surgeons should not have operated on Ms Dinoire, the story goes, because of her psychological susceptibility, evidenced by her attempt to commit suicide. The success of the world's first operation was compromised by the action of her surgeons to go ahead with the transplant on a candidate who falls outside the scope of an ideal patient. She would not necessarily have the psychological wherewithal to make someone else's face her own, to cope with the stress of the media, the long road to recovery in an immune-system-suppressed world. Ongoing assessments of the field continue to refer to the case as an anomaly that perhaps should not be included because of the negligence on the part of the preceding team

DOI: 10.1057/9781137452726.0005

to select a more apt patient. In doing so, they mark her difficulties as physician-initiated failures. This relegates her struggles to the ethical sidelines of the field.

Ms Diniore has resisted the attempts of her presiding medical team to have her relate to her new flesh in a certain way. I see this resistance as part of an ongoing ethical act designed to bring into being a new form of life. Her struggles recounted above, struggles of disgust and repulsion, have been accompanied by a profound attempt to let her donor continue to live on with her. While the team of psychologists and psychiatrists overseeing Ms Diniore want her to appropriate the inside of her mouth, nose and chin directly into her sense of self, to make them her own, she continues to forge her own way of existing with the body parts of another person. Underscoring this ethic is a sense of connectedness between donor and recipient, in which the latter wishes the former to live on through her:

> I found out from the journalist that she had committed suicide. It was one month after the transplant. Somewhere, we were connecting. Two attempts saved one of us. It is strange to know that she wanted to die like me. Strange to know it was she who saved me ... I appropriated it insofar as it is I that managed to make it move. [But] forget it? No. I don't want to and I won't do it. She exists in me. She is and she will always be a part of me. She is my saviour, like a twin sister. This is the only way it can be ... For the doctors, it [the transplant] should be integrated, but it is still not mine. I don't know. It is hard to explain. What I know is that I don't want anyone to damage it, given what the donor offered me. It is too big. Now I am fighting for two![26]

This form of non-compliance hints at a sense of responsibility that Ms Dinoire has toward herself and her donor. Their mutual suicide attempts, one successful the other not, brought them together in the most intimate way. The donor has saved Ms Dinoire, and now it is her responsibility to 'fight for two.' For the overseeing experts, this represents a failure of sorts, as their patient has decided to live outside the criteria established to assess a normal life. While the quality of the life Ms Dinoire lives may not correspond with the dominant institutionalized indicators, in declining to appropriate the graft she makes known the value that she has for it. The ethics of the act lies in her view of an appropriate relationship between donor and recipient in face transplantation. In refusing to 'appropriate' the graft and laying out the reasons why, Ms Dinoire is establishing a set of criteria about how to act in her circumstances.

The challenge therein lies in how to incorporate such individual ethical moves into the broader frameworks of institutional ethics. How can and

DOI: 10.1057/9781137452726.0005

should such practice come to count? Face transplant ethics continues to focus on the practice of surgeons alongside 'ideal patients.' The latter category of person is central to the field because it allows for the operation to take place given the current techno-medical constraints – especially the reliance on harmful immunosuppressant drugs. If the operation is considered anything but a success, as the initial case demonstrates, the failure is attributed to the surgeons for selecting a non-ideal patient. The ethical character of patients has come to hinge on taking pills and living a healthy lifestyle. Lost is a view of ethics as daily struggles to live with new flesh, to figure out where responsibility falls and to make sense of and be in a world in which you are more than one person.

Anthropologists continue to move toward an analysis that breaks down the barrier between bioethics and daily medical life. A central critique of institutional ethics is that it fails to account for the idiosyncrasies of the 'local moral worlds' in which patients and families live.[27] Another take is that bioethics has a role in the very production of everyday medical decision-making. Sometimes, the principles of bioethics may lead to new forms of inaction. Here, they have been central to the production of a fetishized type of patient, one whose suffering allows for medical experiment. The problem that I see is that there is not much room to move when it comes to the tinkering involved in working out how to live post operation; deviation results in the label of deviant, unethical patient-self.

Remaking the ethical: A Normative comment on the future of the field

In arguing for an understanding of 'the face as an organ,' Maria Siemionow suggests that if a comparable sized auto-graft were to be taken from a patient's back, for example, it would be like a 'cardboard mask.'[28] Her argument can itself be viewed as an ethical act, a practice designed to bring an emerging understanding and with it a new imperative into the world: Face transplant surgery is as important as other established forms of organ transplantation because it allows patients to eat, breath, see the world all the while having a positive impact on their blood pressure and other vital aspects of their being.[29] This understanding has since been picked up and used by Dr Siemionow's colleagues to argue for such things as the ethicality of offering the operation to blind patients.

DOI: 10.1057/9781137452726.0005

In the book *Designs on Nature*, Sheila Jasanoff offers the term civic epistemologies to reference nation-specific ways of making sense of new technoscientific practices.[30] By now it should be clear that national ethical reasoning surrounding face transplant surgery has to a large extent been uniform. There has been no outright institutional blocking of the procedure; and the criteria established by various working parties, whether of the Royal College of Surgeons or specific medical teams, resembles more than differs from those of their international colleagues. Indeed, the guidelines, thoughts, and practices delineated by the various parties involved in these early years of face transplantation have been directly borrowed and appropriated by established and new transplant teams alike. The more guidelines are used and applied, the general story goes, the more difficult it becomes to change track and establish a new criteria that represent the reality of the operation and the forms of life it engenders.

Two groups of people are front and centre in this new medical enterprise. Plastic surgeons and their physician colleagues continue to prime the way, to establish what counts and doesn't count as 'good' and 'successful' in this new field. Patients are experimental subjects, people whose misfortune has provided the physical grounds for controversial medical treatment; their bodies, their minds, damaged but healthy enough – it is hoped – to withstand the difficulties of living with the face of another and let the operation take place. The operation pulls these two groups into an integral relationship, a relationship of life and death where the former term denotes anything but a fixed category. Their stories provide a glimpse into a world where institutional ethics and 'ordinary ethics' are constantly on the make, and in doing so alerts the anthropologist bystander to the need to account for the ways in which the extra- and the ordinary rub up against each other in the now everyday existence of a few experimental lives.

Thinking about the words and work of face transplant surgeons requires a mode of inquiry that teases out the space between the mundane and remarkable in which ethical action often takes place. It also poses a challenge to anthropologists concerned with entering into a dialogue with the medico-scientific community. Social studies of science and technology continue to employ ethnographic insights to show how the coming together of various discourses, techniques, and actors often results in the constitution of specific scenarios as ethical problems. These works examine how institutional actors negotiate 'moral landscapes'[31]

DOI: 10.1057/9781137452726.0005

and get caught in the work of 'ethical plateaus', those spaces 'where multiple technologies interact to create a complex terrain or topology of perception and decision making.'[32] The work of face transplant surgeons in many ways echoes these accounts in which ethics is portrayed in terms of negotiation and response, to a melding of both self and environment through tactical action; vis-à-vis the environments in which they are situated, they have produced new criteria for evaluation, called on the State to respond to technoscientific change, and remade themselves through expertise.

To write about face transplant surgery is to reflect on a yet unfinished biomedical enterprise. Though, things, no matter how messy, are being stabilized. This piece of writing should serve as a warning for new technologies of person-making in this world. The various psychological evaluations that patients are subjected to before and after the operation, evaluations to test their psychological shape and acumen, must be remade in order to account for the very novelty of the operation and the modes of existence that it brings about. The very 'instability' that threatens the field, instability displayed by the likes of Ms Dinoire, should be viewed as a crucial tool for figuring out what should count today.

In response to the eighteen-month postoperative report published by Ms Dinoire's surgeons, prominent bioethicist Karen Mashke was critical of (1) a lack of established objective criteria for measuring success, (2) a blurring of the lines between research and innovation, and (3) a lack of postoperative psychological evaluation. In doing so she suggests that it is for the 'transplant community' to agree on the measures for evaluating 'success.'[33] This community, I propose, must include present and potential patients alike. The stakes are too high and the field too new for their expertise to go unaccounted for as important decisions are made regarding how they should live their lives.

Face transplantation calls into question just what counts as 'ethics' and how to integrate the experiences and responses patients and their loved ones have with the operation. A number of scholars from the social sciences have noted that bioethics is often involved and invoked in such a fashion that it justifies or obscures what would otherwise be seen as unethical conduct.[34] Above I mentioned the attempts of anthropologists to forge a conceptual ground between the space of 'high ethics' and the everyday moral deliberation of persons confronted with novel conundrums of how to live. Here I am calling for a new ethics, a new bioethics of face transplantation, one that is not oppositional and polarizing but

DOI: 10.1057/9781137452726.0005

that can be constructed in the space between patient experience and established (psychological) practices of meaning making. What counts as and ways of evaluating 'success' need to be constantly revised as patients continue to figure out how to live post-surgery.

The question is: How can patient experience qua ethics be incorporated into institutional understandings and modes of decision-making? An analogous issue has existed for some time in the domain of patient expertise in medicine more broadly. Studying patient led research projects has allowed analysis of the ways concerned groups are able to influence medical practice and thus 'democratize' technoscience.[35] In a recent address, Jeanette Pols argues that to date however most attempts to account for the place of patient knowledge in broader decision and knowledge making processes has run the risk of, ironically, further demarcating the domains of lay and expert, humanities and medicine. Tackling this issue, she seeks to articulate the object of study, 'the particular knowledge that patients use and develop in their daily practices in order to live with their disease, and how this knowledge relates to forms of medical knowledge.'[36]

Patient knowledge, for Pols, is primarily practical and is coproduced at the intersection of medicine and the sufferer's life conditions:

> [P]atients and people with disabilities develop knowledge and techniques to interpret, appreciate, and shape their daily lives with disease in a good way. 'Good' here is a matter of tinkering and weighing, of coordinating and translating knowledge, technologies, and advice from various sources, including medical practices and technologies. From an epistemological point of view, it is a 'messy' knowledge, involving many different techniques, values, and materials.

Here, the relationship between ethics and knowledge is both intimate and on the make. Pol provides a channel through which, I argue, everyday ethics as how to live can be transported into medical practice. By paying attention to the actions and reactions of patients, comprehending their ways of knowing, and thinking about why they share certain understandings and not others, we might begin to sketch out a mode of 'patient knowledge' that can be used in clinical decision-making. Ethics as figuring out how to act permeates everyday medical practice; likewise, scientific knowledge production is inseparable from understandings of what counts as good practice.[37]

Any assessment of the state of the field and how to protect patients requires that their struggles be regarded as ethical acts comparable to

DOI: 10.1057/9781137452726.0005

those in other medical realms – to the messiness involved in figuring out how to care for yourself and those around you. Delineating guidelines for assessing operation outcomes requires acknowledging the idiosyncrasies of patient experience as people like Ms Dinoire experiment with how to live – how to see and care for themselves – post face transplant surgery. Rather than viewing her difficulties as simply the outcomes of bad decision-making, a position that threatens to marginalise her patient knowledge, they should be seen as providing valuable insights into the affective dimensions of the operation. This involves a realisation that the surgery is doing more than simply reconstructing the anatomy and social life of patients with severe deformities; it is resulting in new ways of being a person.

Notes

1 Ad Hoc Committee of the Harvard Medical School to Examine the Definition of Brain Death (1968). 'A Definition of Irreversible Coma.' *JAMA*, 205(6), 85–88.

2 Giacomini, M. (1997). 'A Change of Health and a Change of Mind? Technology and the Redefinition of Death in 1968.' *Social Science and Medicine*, 44(10), 1465–1482.

3 Ad Hoc Committee of the Harvard Medical School to Examine the Definition of Brain Death, 'A Definition of Irreversible Coma.'

4 Morris, P., A. Bradley, L. Doyal, M. Earley, P. Hagen, M. Milling, & N. Rumsey (2007). 'Face Transplantation: A Review of the Technical, Immunological, Psychological and Clinical Issues with Recommendations for Good Practice.' *Transplantation*, 83(2), 109–129, p. 116.

5 Peralta, E. (2013). 'Falling in Love Again: Face-Transplant Donor's Daughter Meets Recipient.' http://www.npr.org/blogs/thetwo-way/2013/05/03/180892483/falling-in-love-again-face-transplant-donors-daughter-meets-recipient (Accessed 17.09.13), emphasis added.

6 Sharp, L. (2006). *Strange Harvest: Organ Transplants, Denatured Bodies, and the Transformed Self.* Berkeley: University Of California Press.

7 Tarleton, C. (2013). *Overcome: Burned, Blinded and Blessed.* High Land Park, IL: Round Table Press, p. 262.

8 Allan, P. (2007). 'Isabelle Dinoire "May Never Kiss Again".' http://www.telegraph.co.uk/news/worldnews/1564787/isabelle-dinoire-may-never-kiss-again.html (Accessed 19.09.13).

9 For example, Latour, B. (1992). 'Where Are the Missing Masses? The Sociology of a Few Mundane Artifacts.' In W.E. Bijker and J. Law (eds)

DOI: 10.1057/9781137452726.0005

Shaping Technology/Building Society, pp. 225–258. Cambridge, MA: MIT Press.

10 Carty, M., E. Bueno, L. Lehmann, & B. Pomahač (2012). 'A Position Paper in Support of Face Transplantation in the Blind.' *Plastic and Reconstructive Surgery*, 130(2), 319–324, p. 319.

11 Ibid., p. 121.

12 Edwards, J. & D. Mathews (2011). 'Facial Transplantation: A Review of Ethics, Progress, and Future Targets.' *Transplant Research and Risk Management*, 3, 113–125, p. 121.

13 Ibid.

14 Coffman, K., C. Gordon, & M. Siemionow (2010). 'Psychological Outcomes with Face Transplantation: Overview and Case Report.' *Current Opinion in Organ Transplantation*, 15(2), 236–240, p. 258.

15 Ibid., p. 259.

16 Rajan, K. (2005). *Biocapital: The Constitution of Post-Genomic Life*. Durham, NC: Duke University Press.

17 Edwards, J. & D. Mathews (2011). 'Facial Transplantation: A Review of Ethics, Progress, and Future Targets,' p. 121.

18 Carty, M., E. Bueno, L. Lehmann, & B. Pomahač (2012). 'A Position Paper in Support of Face Transplantation in the Blind.'

19 Jackson, M. (2013). *Wherewithal of Life: Ethics, Migration, and the Question of Well-Being*. Berkeley: University Of California Press, p. 13.

20 Lambek, M. (2010). *Ordinary Ethics: Anthropology, Language, and Action*. New York: Fordham University Press, p. 62.

21 Siemionow, M. (2009). *Face to Face: My Quest to Perform the First Full Face Transplant*. New York: Kaplan Publishing, p. 26.

22 Siemionow, M. & E. Sonmez (2011). 'Face as an Organ: The Functional Anatomy of the Face.' In M. Siemionow (ed.) *The Know-How of Face Transplantation*, pp. 3–10. London: Springer, p. 4.

23 Ibid., p. 6.

24 Siemionow, M., B. Gharb, & A. Rampazzo (2011). 'The Face as a Sensory Organ.' In M. Siemionow (ed.) *The Know-How of Face Transplantation*, pp. 11–23. London: Springer, p. 13.

25 Siemionow. *Face to Face*.

26 Cited in Lafrance, M. (2010). '"She Exists within Me": Subjectivity, Embodiment, and the World's First Face Transplant.' in T. Rudge and D. Holmes (eds) *Abjectly Boundless: Boundaries, Bodies and Health Work*, pp. 147–162. London: Ashgate.

27 Kleinman, A. (1995). *Writing at the Margin: Discourse between Anthropology and Medicine*. Berkeley: University of California Press.

28 Siemionow. *Face to Face*.

29 Siemionow, M. & E. Sonmez. 'Face as an Organ.'

DOI: 10.1057/9781137452726.0005

30   Jasanoff, S. (2006). *Designs On Nature*. Princeton: Princeton University Press.

31   Helgason, A. & G. Palsson (1997). 'Contested Commodities: The Moral Landscape of Modernist Regimes.' *Journal of the Royal Anthropological Institute*, 3(3), 451–471.

32   Fortun, K. & M. Fortun (2005). 'Scientific Imaginaries and Ethical Plateaus in Contemporary US Toxicology.' *American Anthropologist*, 107(1), 43–54, p. 47.

33   Mashke, K. & T. Murray (2007). 'Doctors Report on First Partial Human Face Transplant.' *Bioethics Responder*. www.thehastingscenter.org/news/detail.aspx?id=1650 (Accessed 12.03.2013).

34   For example, Rosenberg, C. (1999). 'Meanings, Policies, and Medicine: On the Bioethical Enterprise and History.' *Daedalus*, 128(4), 27–46; Scheper-Hughes, N. (2005). 'The Last Commodity: Post-Human Ethics and the Global Traffic in "Fresh" Organs.' In A. Ong and S. Collier (eds) *Global Assemblages: Technology, Politics, and Ethics as Anthropological Problems*. Chicago: Wiley-Blackwell, pp. 145–167; Petryna, A. (2009). *When Experiments Travel: Clinical Trials and the Global Search for Human Subjects*. Princeton: Princeton University Press.

35   Callon, M. & V. Rabeharisoa (2003). 'Research "In the Wild" And the Shaping of New Social Identities.' *Technology in Society*, 25(2), 193–204.

36   Pols, J. (2013). 'Knowing Patients: Turning Patient Knowledge into Science.' *Science, Technology & Human Values*, 39(1), 73–97.

37   Fortun, K. & M. Fortun (2005). 'Scientific Imaginaries and Ethical Plateaus in Contemporary Us Toxicology.'

DOI: 10.1057/9781137452726.0005

4
Constituting a Field

Abstract: *What is at stake in emerging medical fields is not only the lives of patients, but also the very ways in which state institutions, surgeons, and families make sense of rights, claims for inclusion, and life itself. Face transplantation is still very much a medical field on the make and so provides a useful case for thinking about how new instantiations of rights and responsibilities, health and healing emerge alongside novel technologies. In concluding his book, I show that central to the emergence of face transplantation has been the stabilization, both political and medical, of a particular kind of human subject. The production of persons in face transplantation continues to happen in tandem with the making of institutions, epistemic formations, and understandings of ethical practice. The operation thus challenges us to pay attention to the 'constitutional' dimensions of science and technology, to the central role they play in the ordering of contemporary society.*

Taylor-Alexander, Samuel. *On Face Transplantation: Life and Ethics in Experimental Biomedicine.* Basingstoke: Palgrave Macmillan, 2014. DOI: 10.1057/9781137452726.0006.

The story of face transplantation that I have offered has revealed as inherently political the ideological formations that underscore the practice and demonstrated the lengths that surgeons will go to as they seek to gain the resources needed to perform the operations they desire for their patients (and themselves). This desire exists alongside constantly shifting understandings of care and necessary practice. As these understandings shift, so do treatment methods and experiences of illness. There is a need to rethink institutional approaches to evaluating operative outcomes in contexts where such a shifting of ground takes place. Aside from making minor modifications to existing methods of psychological evaluation, this is yet to take place in the realm of face transplantation. More sophisticated approaches are needed so that the medical community can better grasp what it means to be a patient in this experimental field. This doesn't mean that these professionals are not responding to the experience of patients – far from it. This community of surgeons, ethicists, psychologists and the like are continually looking to patient outcomes in order to make available and modify their unique treatment methods. For the most part, though, these outcomes are produced by established ways of doing things and so the picture these professionals have is limited.

I began this essay by posing a question: How are understandings of 'life' being reworked with the emergence of new technologies? The inverted commas that surround the word 'life' serve to highlight the fluid and multiple things that the term can refer to: molecular, ethical, social, psychological, political. Medical anthropologist Byron Good argues that illness at once forms part of a lived experience, is an object of political and therapeutic attention, and is a physiological condition.[1] The heterology, or multiplicity, of illness has further been explored by sociologist Annemarie Mol, who through her analysis of atherosclerosis demonstrates the ways in which 'medicine enacts the objects of its concern and treatment,' that is, it 'attunes to, interacts with, and shapes its objects in its various and varied practices.'[2] For Mol, the multiplicity of illness is constituted in relation to the various environments in which it is examined and experienced. Where Good points us to the political factors involved in constructions of illness, Mol talks of the very making of illness in its different settings. This case at hand demonstrates the need to pay attention to the space between these two spheres, between social and clinical order.

Face transplantation is still very much a medical field on the make and so provides a useful case for thinking about how understandings

DOI: 10.1057/9781137452726.0006

of life shift with the introduction of new technologies. Involved in its making are a number of various actors: people, ideals, institutions, laws, nation-states, technologies, human-tissue. One of these actors is a discursive, medico-political construct: the 'ideal patient'. The existence of this 'person' is demonstrative of the categorisation and classification that often takes place in the context of emergent techno-medical fields. The philosopher Ian Hacking has spoken of classification in medicine as a process of 'making up people' whereby the very subjective being of patients is produced anew as they become the object of analysis and intervention. As new categories of person, such as the autistic, are established, the very character of these people changes with their diagnosis and treatment.[3] As with the practices at the centre of Hacking's analysis, classification in face transplantation simultaneously stabilises and shifts people as institutional beings. These people are both real and imagined; and they are at the centre of tactical efforts to produce an account in which the operation is seen as providing an appropriate form of life.

This is the power and the importance of the 'ideal patient' of face transplantation. This notion is normative. Like the very idea of the (ab)normality upon which the entire field of reconstructive surgery is built, it delineates a type of person in need of a specific form of medical intervention. As we have seen, this patient-in-need was produced at the intersection of surgical advancements and technical limitations and patient-doctor relationships – the very notions of health and the human self that underscore biomedical practice. Embedded within this construct is an account of an appropriate form of life, a life worth living and worth risking death to achieve. As the story of Escobar and his surgical carers reminds us, the 'ideal patient' is part of a larger feedback loop. This loop began as much as anywhere in the need to satisfy institutional requirements; and it resulted in a shifting of relationships and responsibilities between patients, doctors, donors and the institutions in which they are embedded. It is not just people being made, here, but institutional forms and bureaucratic apparatuses themselves.

The constitution of persons as suitable face transplant subjects is central to the ongoing making of the field. This is what makes the operation a political and medical experiment. The lives of patients like Isabelle Dinoire and Carmen Tarleton, along with their joy, suffering, and disgust, are vital parts of this experiment. Though, their experiences and knowledge are obscured by the bureaucratic technologies that require

DOI: 10.1057/9781137452726.0006

not subjective and personal but *objective and standardized* accounts of a messy, constantly shifting world. Standards and standardization are the topic of increasing attention from social scientists. More so than ever before, the growth of statistical measures is central to the work of institutions around the globe.[4] A danger of this trend is that numerical assessments are produced according to pre-existing value systems and are unable to represent the social and political complexity surrounding, for example, why some doctors have better surgical outcomes than others[5] or the worth of organisations in the charity sector.[6] In these scenarios, ethics and morals come to exist in the practice of quantification itself, so that claims to be an ethical charity or a good doctor are made through the production of performance indicators.

Face transplantation reminds us that it is difficult to see medicine and politics as anything but interconnected and overlapping when we stop to take a good look at emerging treatment methods. This overlap is most visible in the work of medical bureaucracy. As we saw above, this bureaucratization occurred post-WWII largely in response to the atrocities committed by Nazi doctors who made Jewish prisoners the locus of inhumane biomedical experimentation. The resulting declarations and bioethical principles sought to protect both patient populations and the reputation of biomedicine. In doing so, they further embedded within the domain a particular, western notion of the *individual* person – a rights-bearing subject, responsible for both their own wellbeing and that of the society in which they live.

Some fifty years later when surgeons began to consider the plausibility of face transplantation as a therapeutic option they had to demonstrate to the institutions in which they (and their peers) worked that the operation could be performed without moving all those involved into unethical territory. What is significant about how they achieved this lies in their production of a standard account of a person: the ideal patient. This is a person who can be analysed and interpreted; who can travel across national boundaries, between institutions and their various spaces while maintaining a certain element of ontological and epistemic consistency. The make up of the person/self, embodied in surgical protocols and ethical appraisals, is able to move between different settings without losing its essential characteristics.

Here, I am pointing to the requirements and effects of medical bureaucracy. The practices that institutions seek to make sense of and govern are re-ordered according to established modes of producing and

DOI: 10.1057/9781137452726.0006

interpreting information. What they come to appraise is a product forged by the methods of surveillance and the values of those performing it. Doctors must enter into a conversation with institutional bodies in order to gain the permission necessary to perform experimental procedures. Meanwhile this conversation is framed from the outset by established (economically influenced) modes of decision-making, by QALYS and risk-benefit analysis. The importance of QALYS and other measurement systems is central to the making of face transplantation because they are based on the same understandings of the individual self upon which plastic surgery writ large has gained its legitimacy as a medical practice. This means that the account of the operation accessible to regulators and the like is devoid of the newfound complexity that it has introduced into the lives of patients.

In order for people to be able to make decisions in these contexts they must first be able to make sense of the information they have before them. Messy patient accounts will not do here; rather information needs to conform to established ways of appraising and ordering the world. The categorisation of patients, and the related presenting of numerical and statistical reports that quantify the impact and importance of face transplantation for their lives, is central in the solidification and shifting of patients like Ms Dinoire and Ms Tarleton. It allows them to be enrolled as actors in the emerging medical field, to be operated upon and remade anew, while keeping stable the ideas of the person that now underscores the field. These are ideas relating both to what constitutes an appropriate form of life (the 'normal', 'healthy' citizen) and to the person/self that is produced through the reasoning of bureaucratic bodies confronted with what perhaps still is an 'ethical nightmare' (the ideal patient).

The production of persons in face transplantation continues to happen in tandem with the making of institutions, epistemic formations, and understandings of ethical practice. Underscoring this making is a number of norms and accepted forms of institutional practice, from bioethical reason to long established medical techniques and national laws. Sheila Jasanoff has termed 'constitutional' those moments forged at the meeting of technoscience and politics that bring profound shifts in citizen-state relations. Challenging constitutional theorists, she writes:

> Given the centrality of scientific knowledge and technological artefacts in contemporary life, it is reasonable to think that the basic ordering

DOI: 10.1057/9781137452726.0006

commitments of modern societies will be found not only in legal texts, but also, tacitly expressed, the very organisation of life around the products of human ingenuity and knowledge … Order may emerge not merely, or even mainly, when positive law bestows it or a court affirms it, but also when people assume that they have the capacity and the right to change their behaviour in fundamental ways, and act accordingly.[7]

The constitutional dimensions of face transplantation are most clearly visible in the Mexican case, where lawyer Octavio Madrid Mata called on the state to fulfil its legal obligations and make available face transplantation to the nation's citizens.

It was the very failure of the Mexican health care system that resulted in face transplantation becoming the focus of medical and political deliberation in Mexico. The severe loss of tissue and threat to Escobar's overall health resulted from a foundering on the part of the State to ensure his constitutional right – to protect his health. Face transplantation in this context emerged as a site in which the relationship between the nation-state, medical science, and citizenship were reimagined as doctors and lawyers came together to lay the ground necessary for treating Escobar in a world where the operation was already a reality: It was his condition, his state of health and suffering that conformed to a number bureaucratically defined measures that brought this relationship into being. It was not only Escobar's face that was under construction, but also the image of Mexico itself as it looked outwardly towards the world, towards establishing itself as an international actor on the cutting edge of biomedicine. Yet, ironically, Escobar was in effect left faceless through a transmutation that configured his treatment options as a nation-making project. His surgeons were forced to use traditional reconstructive methods, leaving him all but unable to eat and speak, to treat a condition that spurred from the original denial he experienced when seeking treatment for an abscessed tooth. And even though the operation has yet to be performed in Mexico, it has altered the relationships between individuals and institutions, resulted in new forms of responsibility and given birth to new human and bureaucratic bodies. Here, the heterology of illness exists in both its clinical and political enactments. Ethical review boards, patients, health agencies, and medical professionals enact facial disfigurement as they partake in the production of new medico-political understandings of what counts as a good or appropriate life.

DOI: 10.1057/9781137452726.0006

Notes

1   Good, B. (1994). *Medicine, Rationality and Experience: An Anthropological Perspective.* Cambridge, UK: Cambridge University Press, chapter 7.
2   Mol, A. (2002). *The Body Multiple: Ontology in Medical Practice.* Durham, NC: Duke University Press, p. vii.
3   Hacking, I. (2005). 'Making Up People.' *London Review of Books,* 28(16), 23–26.
4   Davis, K., B. Kingsbury, & S. Merry (2012). 'Indicators as a Technology of Global Governance.' *Law and Society Review,* 46(1), 71–104.
5   Taylor-Alexander, S. (Under Review). 'Ethics in Numbers: Cleft Audit in Mexico and Beyond.' *Medical Anthropology Quarterly.*
6   Merry, S. E. (2011). 'Measuring the World.' *Current Anthropology,* 52(S3), 83–95.
7   Jasanoff, S. (2003). 'In a Constitutional Moment: Science and Social Order at the Millennium.' In B. Joerges and H. Nowotny (eds) *Social Studies of Science and Technology: Looking Back, Ahead,* pp. 155–180. Dordrecht: Kluwer Academic Publishers, p. 161.

DOI: 10.1057/9781137452726.0006

Postscript

The broader significance of face transplantation is still to be felt. A number of questions can be raised surrounding how it is shaping the experience of patients who are unable to undergo the operation: are they likely to be less satisfied with the outcomes of the treatment they do receive and follow the calls of reconstructive surgeons and test their rights in the politico-legal realm? The present, and its anecdotal accounts, offers us a few clues. Joel McNicholls, a twenty year-old Englishman, recently spent five months in the University Hospital of Southern Manchester for the treatment of facial burns and disfigurement resulting from the crash of a small passenger plane. Through his solicitor, he has filed a writ in the British High Court and is suing the aircrafts operator 'Cheshire Flying Services' for damages and compensation, including finances to fund a possible face transplant.[1] Mr McNicholls is enacting his illness and his rights through the courts, using the law as a possible means to gain access to the care that he desires in a world of ever changing expectations of medical treatment. The trauma that is now part of Joel McNicholls daily life and sense of self is of a different variety to what it would be in a present where face transplantation did not exist. Reflecting on his plight, I can't help but think of the numerous cases around the globe where people struggle in courtrooms to gain access to 'essential' medicines; and the new role of the judiciary in the growth of biologically mediated forms of citizenship.[2]

Medical professionals are already starting to ask who will fund the operation once/if it moves from experimental

DOI: 10.1057/9781137452726.0007

medicine to standard treatment. Unexplored in this essay is the financial cost of face transplantation. Proponents of the operation, like Maria Siemionow, have been quick to produce models that show as similar its cost when compared to traditional reconstructive operations.[3] Though, her professional colleagues remind us that the many interest groups and charitable organisations interested in the procedure have supplied the necessary financial resources that have so far funded the field. Whereas the concerns surrounding the field to date have largely focused on whether the operation can be performed successfully this does not mean that in the future surgeons will be able to transplant composite facial tissue: 'While initial examination of cost-effectiveness may argue for or against facial transplantation, the final determinant of whether this procedure is funded will be how society views the benefits of the procedure,' argue surgeons Edwards and Mathews.[4] What society is, in their reckoning, is not clear: judges, the public, insurance companies, patient organisations? What is clear is that these views are beginning to be brought into play, measured and formulated as the public,[5] plastic surgeons and institutions are asked to reflect on the operation and its benefits and dangers.

At the same time a number of therapeutic techniques are being developed that could displace the very relevance of face transplantation for patients with severe craniofacial conditions. I was recently reminded of these alternatives when a colleague of mine asked me to a review the video of a TED talk that she was considering showing before my guest lecture in her undergraduate anthropology class. In the video Iain Hutchinson, founder of the Saving Faces Foundation, provides an overview of his methods of near total facial reconstruction.[6] A renowned critic of face transplantation, he rehearses the same arguments that I introduced at the beginning of this book – the benefits of a new life versus the risks of death resulting from the side effects of immunosuppression and other complications – before offering an alternative to face transplantation: tissue engineering. The idea of fabricating living tissue to aid facial reconstruction has been around since the turn of the millennium. It involves the production of biodegradable material that can be injected with a patients own stem-cells, which as they grow replace the original structure with what essentially amounts to a living copy of the person's facial tissue.[7] Whether or not this method of facial reconstruction will eventually exist alongside or replace face transplantation, it is increasingly finding a place in the voice of critics of the experimental

DOI: 10.1057/9781137452726.0007

biomedical practice. It also reminds us that in evaluating face transplantation, we need to think about not only what is happening in the field but also consider broader developments in clinical and laboratory medicine.

When I read the accounts of patients that have undergone the operation and view the before and after pictures that circulate the internet, I cannot help but feel moved by their stories. When someone asks me whether I think face transplantation is a good or bad thing, I struggle to find an answer but try my best to provide a response that captures the complexity of the field. I tell them of the dangers of immunosuppression, the stories of Carmen Tarleton and Isabelle Dinoire and Connie Culp, of the possibilities of tissue engineering. I hope that I have provided you, the reader, with a response that captures the complexity of facial transplantation as a form of experimental biomedicine. As with other scientific and technical developments, it provides both the concerned bystander and the involved expert a chance to reflect on what it means to be a person today; it helps us to better understand how notions of self and rights and normality shift in a world that is increasingly made up of new, experimental forms of human life.

Notes

1   BBC (2003). 'Salford Plane Crash Survivor Seeks Damages to Pay for Face Transplant.' http://www.bbc.com/news/uk-england-manchester-22544885 (Accessed 06.03.2014).

2   See for example: Biehl, J. (2013). 'The Judicialization of Biopolitics: Claiming the Right to Pharmaceuticals in Brazilian Courts.' *American Ethnologist*, 40(3), 419–436; Trundle, C. and B. Scott (2013). 'Elusive Genes: Nuclear Test Veterans' Experiences of Genetic Citizenship and Biomedical Refusal.' *Medical Anthropology*, 32(6), 501–517.

3   Siemionow, M., B. Gharb, & A. Rampazzo (2011). 'Cost Analysis of Conventional Facial Reconstruction Procedures Followed By Face Transplantation.' *American Journal of Transplantation*, 11(2), 379–385.

4   Edwards, J. and D. Mathews (2011). '*Facial Transplantation: A Review of Ethics, Progress, and Future Targets,*' p. 121.

5   Clarke, A., J. Simmons, & P. White (2006). 'Attitudes to Face Transplantation: Results of a Public Engagement Exercise at the Royal Society Summer Science Exhibition.' *Journal of Burn Care & Research*, 27(3), 394–398; Tan, W., A. Patel, P. Taub, J. Lampert, G. Xipoleas, G. Santiago, L. Silver, H. Sheriff, T-S. Lin,

DOI: 10.1057/9781137452726.0007

R. Cooter, F. Diogo, B. Salazaard, B. J. Kim, Y. Lee and R. Ogawa (2011). 'Cultural Perspectives in Facial Allotransplantation.' *Eplasty*, 12, 344–353.

6 Hutchinson, I. (2010). 'Saving Faces: A Plastic Surgeons Craft.' http://www.ted.com/talks/iain_hutchison_saving_faces?utm_source=email&source=email&utmmedium=social&utm_campaign=ios-share (Accessed 07.03.2014).

7 Vacanti, J. P. and R. Langer (1999). 'Tissue Engineering: The Design and Fabrication of Living Replacement Devices for Surgical Reconstruction and Transplantation.' *The Lancet*, 354(S), 32–34.

DOI: 10.1057/9781137452726.0007

Bibliography

Ad Hoc Committee of the Harvard Medical School
to Examine the Definition of Brain Death. (1968).
'A Definition of Irreversible Coma.' *Journal of the
American Medical Association*, 205(6), 85–88.

Agamben, G. (2000). *Means without End: Notes on
Politics*. Minneapolis: University of Minnesota Press.

Allan, P. (2007). 'Isabelle Dinoire "May Never Kiss Again".'
http://www.telegraph.co.uk/news/worldnews/1564787/
isabelle-dinoire-may-never-kiss-again.html (accessed
19.09.2013).

BBC. (2003). 'Salford Plane Crash Survivor Seeks
Damages to Pay for Face Transplant.' http://www.bbc.
com/news/uk-england-manchester-22544885 (accessed
06.03.2014).

Biehl, J. (2013). 'The Judicialization of Biopolitics:
Claiming the Right to Pharmaceuticals in Brazilian
Courts.' *American Ethnologist*, 40(3), 419–436.

Callon, M. & Rabeharisoa, V. (2003). 'Research "in
the Wild" and the Shaping of New Social Identities.'
Technology in Society, 25(2), 193–204.

Camfield, L. (2002). *Measuring Quality of Life in Dystonia:
An Ethnography of Contested Representations*. Phd
Thesis: Goldsmiths College.

Carty, M.J., Bueno, E.M., Lehmann, L.S. & Pomahac,
B. (2012). 'A Position Paper in Support of Face
Transplantation in the Blind.' *Plastic and Reconstructive
Surgery*, 130(2), 319–324.

Clarke, A., Simmons, J. & White, P. (2006). 'Attitudes to
Face Transplantation: Results of a Public Engagement

DOI: 10.1057/9781137452726.0008

Exercise at the Royal Society Summer Science Exhibition.' *Journal of Burn Care and Research*, 27(3), 394–398.

Coffman, K.L., Gordon, C. & Siemionow, M. (2010). 'Psychological Outcomes with Face Transplantation: Overview and Case Report.' *Current Opinion in Organ Transplantation*, 15(2), 236–240.

Cohen, L. (2005). 'Operability, Bioavailability and Exception.' in A. Ong and S. Collier (eds) *Global Assemblages: Technology, Politics, and Ethics as Anthropological Problems*, pp. 79–90 Oxford: Wiley-Blackwell.

Davis, K., Kingsbury, B. ,& Merry, S. (2012). 'Indicators as a Technology of Global Governance.' *Law and Society Review* 46(1), 71–104.

Edwards, J. & Mathews, D. (2011). 'Facial Transplantation: A Review of Ethics, Progress, and Future Targets.' *Transplant Research and Risk Management*, 3, 113–125.

Fortun, K. & Fortun, M. (2005). 'Scientific Imaginaries and Ethical Plateaus in Contemporary US Toxicology.' *American Anthropologist*, 107(1), 43–54.

Giacomini, M. (1997). 'A Change of Health and a Change of Mind? Technology and the Redefinition of Death in 1968.' *Social Science and Medicine* 44(10), 1465–1482.

Gilman, S. (1998). *Creating Beauty to Cure the Soul*. Durham: Duke University Press.

Gilman, S. (2000). *Making the Body Beautiful: A Cultural History of Aesthetic Surgery*. Princeton, NJ: Princeton University Press.

Good, B. (1994). *Medicine, Rationality and Experience: An Anthropological Perspective*. Cambridge, UK: Cambridge University Press.

Good, M-J. (2001). 'The Biotechnical Embrace.' *Culture, Medicine and Psychiatry*, 25(4), 395–410.

Hacking, I. (1990). *The Taming of Chance*. Cambridge, UK: Cambridge University Press.

Hacking, I. (2005). 'Making Up People.' *London Review Of Books*. 28(16), 23–26.

Helgason, A. & Palsson, G. (1997). 'Contested Commodities: The Moral Landscape of Modernist Regimes.' *Journal Of The Royal Anthropological Institute*, 3(3), 451–471.

Hutchinson, I. (2010). Saving Faces: A Plastic Surgeon's Craft. http://www.ted.com/talks/iain_hutchison_saving_faces?utm_source=email&source=email&utmmedium=social&utm_campaign=ios-share (Accessed 07.03.2014).

DOI: 10.1057/9781137452726.0008

Infante-Cossio, P., Barrera-Pulido, F., Gomez-Cia, T., Sicilia-Castro, D., Garcia-Perla-Garcia, A., Gacto-Sanchez, P., Hernandez-Guisado, J., Lagares-Borrego, A., Narros-Gimenez, R. & Gonzalez-Padilla, J. (2013). 'Facial Transplantation: A Concise Update.' *Medicina Oral, Patologia Oral Y Cirugia Bucal,* 18(2), 37.

Jackson, M. (2013). *Wherewithal of Life: Ethics, Migration, and the Question of Well-Being.* Berkeley: University of California Press.

Jasanoff, S. (2003). 'In a Constitutional Moment: Science and Social Order at the Millennium.' In B. Joerges and H. Nowotny (eds) *Social Studies of Science and Technology: Looking Back, Ahead,* pp. 155–180. Dordrecht: Kluwer Academic Publishers.

Jasanoff, S. (2006). *Designs on Nature.* Princeton: Princeton University Press.

Joralemon, D. & Fujinaga, K. (1997). 'Studying the Quality of Life after Organ Transplantation: Research Problems and Solutions.' *Social Science and Medicine* 44(9), 1259–1269.

Kleinman, A. (1995). *Writing at the Margin: Discourse between Anthropology and Medicine.* Berkeley: University of California Press.

Lafrance, M. (2010). '"She Exists within Me": Subjectivity, Embodiment, and the World's First Face Transplant.' in T. Rudge and D. Holmes (eds) *Abjectly Boundless: Boundaries, Bodies and Health Work,* pp. 147–162. London: Ashgate.

Latour, B. (1992). 'Where Are the Missing Masses? "the Sociology of a Few Mundane Artifacts."' in W.E. Bijker and J. Law (eds) *Shaping Technology/Building Society,* pp. 225–258. Cambridge; MA: MIT Press.

Lambek, M. (2010). *Ordinary Ethics: Anthropology, Language, and Action.* New York: Fordham University Press

Lock, M. (2001). *Twice Dead: Organ Transplants and the Reinvention of Death.* Berkeley: University of California Press.

Madrid Mata, O. (2007). 'Trasplante de Tejido Facial. Aspectos Jurídicos.' *Cirugia Plastica,* 17(1), 31–48.

Merry, S. E. (2011). 'Measuring the World.' *Current Anthropology,* 52(S3), 83–95.

Mol, A. (2002). *The Body Multiple: Ontology in Medical Practice.* Durham, NC. Duke University Press.

Morris, P., Bradley, A., Doyal, L., Earley, M., Milling, M. & Rumsey, N. (2003). *Facial Transplantation: Working Party Report.* London: Royal College of Surgeons of England.

DOI: 10.1057/9781137452726.0008

Morris, P., Bradley, A., Doyal, L., Earley, M., Hagan, P., Milling, M. & Rumsey, N. (2006). *Facial Transplantation: Working Party Report* (2nd Edition). London: Royal College of Surgeons of England.

Morris, P., Bradley, A., Doyal, L., Earley, M., Hagan, P., Milling, M. & Rumsey, N. (2007). 'Face Transplantation: A Review of the Technical, Immunological, Psychological and Clinical Issues With Recommendations for Good Practice.' *Transplantation* 83(2), 109–129.

Naugler, D. (2009) 'Crossing the Cosmetic/Reconstructive Divide: The Instructive Situation of Breast Reduction Surgery.' in M. Jones and C. Hayes (eds) *Cosmetic Surgery: A Feminist Primer*, pp. 225–238. Farnham, UK: Ashgate.

Peralta, E. (2013). 'Falling in Love Again: Face-Transplant Donor's Daughter Meets Recipient.' http://www.npr.org/blogs/thetwo-way/2013/05/03/180892483/falling-in-love-again-face-transplant-donors-daughter-meets-recipient (Accessed 17.09.2013).

Petryna, A. (2009). *When Experiments Travel: Clinical Trials and the Global Search for Human Subjects*. Princeton: Princeton University Press.

Pigott, R. (1995). 'Aesthetic Reconstructive Surgery.' *British Journal of Plastic Surgery*, 48(5), 338–339.

Pols, J. (2013). 'Knowing Patients: Turning Patient Knowledge into Science.' *Science, Technology and Human Values*, 39(1), 73–97.

Rajan, K. (2005). *Biocapital: The Constitution of Post-Genomic Life*. Durham, NC: Duke University Press.

Rosenberg, C. (1999). 'Meanings, Policies, and Medicine: On the Bioethical Enterprise and History.' *Daedalus*, 128(4), 27–46.

Scheper-Hughes, N. (2005). 'The Last Commodity: Post-Human Ethics and the Global Traffic in "Fresh" Organs.' In A. Ong and S. Collier (eds) *Global Assemblages: Technology, Politics, and Ethics as Anthropological Problems*, pp. 145–167. Chicago: Wiley-Blackwell.

Sharp, L. (2006). *Strange Harvest: Organ Transplants, Denatured Bodies, and the Transformed Self*. Berkeley: University of California Press.

Siemionow, M. (2009). *Face to Face: My Question to Perform the First Full Face Transplant*. New York: Kaplan.

Siemionow, M. & Sonmez, E. (2008). 'Face as an Organ.' *Annals of Plastic Surgery*, 61(3), 345–352.

Siemionow, M., Gharb, B. & Rampazzo, A. (2011). 'The Face as a Sensory Organ.' In M. Seimionow (ed.) *The Know-How of Face Transplantation*, pp. 11–23. London: Springer.

DOI: 10.1057/9781137452726.0008

Siemionow, M., Gatherwright, J., Djohan, R. & Papay, F. (2011). Cost Analysis of Conventional Facial Reconstruction Procedures Followed by Face Transplantation. *American Journal of Transplantation*, 11(2), 379–385.

Slatman, J. & Widdershoven, G. (2010). 'Hand Transplants and Bodily Integrity.' *Body and Society*, 16(3), 69–92.

Strathern, M. (1988). *The Gender of the Gift: Problems with Women and Problems with Society in Melanesia*. Berkeley: University of California Press.

Tan, W. Patel, A., Taub, P., Lampert, J., Xipoleas, G., Santiago, G., Silver, L., Sheriff, H., Lin, T-S., Cooter, R., Diogo, F., Salazaard, B., Kim, B. J., Lee, Y. & Ogawa, R. (2012). 'Cultural Perspectives in Facial Allotransplantation.' *Eplasty*, 12, 344–353.

Tarleton, C. (2013). *Overcome: Burned, Blinded and Blessed*. High Land Park; IL: Round Table Press.

Taylor-Alexander, S. (Under Review). 'Ethics in Numbers: Cleft Audit in Mexico and beyond.' *Medical Anthropology Quarterly*.

The Hastings Centre. (2007). 'Doctors Report on First Partial Human Face Transplant.' *Bioethics Responder*. www.thehastingscenter.org/news/detail.aspx?id=1650 (Accessed 12.03.2013)

The Lancet. (2005). 'The First Face Transplant.' *The Lancet*, 366(9502), p. 1984.

Trundle, C. & Scott, B. (2013). 'Elusive Genes: Nuclear Test Veterans' Experiences of Genetic Citizenship and Biomedical Refusal.' *Medical Anthropology*, 32(6), 501–517.

Vacanti, J. P. & Langer, R. (1999). 'Tissue Engineering: The Design and Fabrication of Living Replacement Devices for Surgical Reconstruction and Transplantation.' *The Lancet*, 354(S32–S34).

Visvanathan, S. (1997). *A Carnival for Science: Essays on Science, Technology, and Development*. Delhi: Oxford University Press.

DOI: 10.1057/9781137452726.0008

Index

DOI: 10.1057/9781137452726.0009

▶ **Network Society and Future Scenarios for a Collaborative Economy**

DOI: 10.1057/9781137406897.0001

Other Palgrave Pivot titles

DOI: 10.1057/9781137406897.0001

palgrave▸**pivot**

Network Society and Future Scenarios for a Collaborative Economy

▶ Vasilis Kostakis
Research Fellow, Tallinn University of Technology, Estonia

and

Michel Bauwens
Founder, P2P Foundation

palgrave
macmillan

DOI: 10.1057/9781137406897.0001

© Vasilis Kostakis and Michel Bauwens 2014

All rights reserved. No reproduction, copy or transmission of this
publication may be made without written permission.

No portion of this publication may be reproduced, copied or transmitted
save with written permission or in accordance with the provisions of the
Copyright, Designs and Patents Act 1988, or under the terms of any licence
permitting limited copying issued by the Copyright Licensing Agency,
Saffron House, 6–10 Kirby Street, London EC1N 8TS.

Any person who does any unauthorized act in relation to this publication
may be liable to criminal prosecution and civil claims for damages.

The authors have asserted their rights to be identified as the author of this work
in accordance with the Copyright, Designs and Patents Act 1988.

First published 2014 by
PALGRAVE MACMILLAN

Palgrave Macmillan in the UK is an imprint of Macmillan Publishers Limited,
registered in England, company number 785998, of Houndmills, Basingstoke,
Hampshire RG21 6XS.

Palgrave Macmillan in the US is a division of St Martin's Press LLC,
175 Fifth Avenue, New York, NY 10010.

Palgrave Macmillan is the global academic imprint of the above companies
and has companies and representatives throughout the world.

Palgrave® and Macmillan® are registered trademarks in the United States,
the United Kingdom, Europe and other countries.

ISBN: 978–1–13740–688–0 EPUB
ISBN: 978–1–13740–689–7 PDF
ISBN: 978–1–13741–506–6 Hardback

A catalogue record for this book is available from the British Library.

A catalog record for this book is available from the Library of Congress.

www.palgrave.com/pivot

DOI: 10.1057/9781137406897

Contents

DOI: 10.1057/9781137406897.0001

List of Figures

Preface

The aim of this book is not to provide yet another critique of capitalism but rather to contribute to the ongoing dialogue for post-capitalist construction, and to discuss how another world could be possible. We build on the idea that peer-to-peer infrastructures are gradually becoming the general conditions of work, economy and society, considering peer production as a social advancement within capitalism but with various post-capitalistic aspects in need of protection, enforcement, stimulation and connection with progressive social movements. Using a four-scenario approach, we attempt to simplify possible outcomes and to explore relevant trajectories of the current techno-economic paradigm within and beyond capitalism. The first part of the book begins with an introduction (Chapters 1 and 2) of the techno-economic paradigm shifts theory, which sees capitalism as a creative destruction process. Such a dynamic, innovation-based understanding of economic and societal development arguably allows for an integral bird's-eye view of future scenarios (Chapter 3) within and beyond the dominant system. Sharing the conviction that the globalized economy is at a critical turning point, we describe the four future scenarios: netarchical capitalism, distributed capitalism, resilient communities and global Commons. Netarchical and distributed capitalism (Chapters 4 and 5) are parts of the wider value mode of cognitive capitalism and form, what we call 'the mixed model of neo-feudal cognitive capitalism' (Chapter 6). On the other hand, the resilient communities (Chapter 7) and the global Commons (Chapter 8) reside in the

DOI: 10.1057/9781137406897.0003

hypothetical model of mature peer production under civic dominance. We postulate that the mature peer production communities pose a sustainable alternative to capital accumulation, that of the circulation of the Commons. Hence, we make some tentative transition proposals toward a Commons-based economy and society for the state, the market and the civic domain (Chapter 9). Finally, we conclude with remarks and suggestions for future actions.

DOI: 10.1057/9781137406897.0003

Acknowledgments

We would like to express our very great appreciation to Christos Giotitsas, Denis Postle, Katarzyna Gajewska, Helene Finidori and Nikos Anastasopoulos for their constructive suggestions during the planning and development of this research work. In addition to this, we are particularly indebted to Vasilis Niaros for his support in the editing of the book as well as in the designing of the figures. Further, we owe gratitude to Wolfgang Drechsler, Nikos Salingaros, Rainer Kattel and Carlota Perez who have been mentoring our work for years now. Moreover, we would like to extend our thanks to Ann Marie and Stacco from Guerrilla Translation! for carefully copy-editing the text; as well as to Christina Brian, Head of Politics & International Studies at Palgrave Macmillan, and Ambra Finotello, editorial assistant, for their constant support, understanding and eagerness. Also, the work of the FLOK Society, a collaborative research effort in Ecuador, has been crucial to develop transition and policy proposals toward a Commons-based knowledge society. Michel Bauwens was the research director of the FLOK society project and Vasilis Kostakis served as an external collaborator. The latter also acknowledges financial support by the 'Challenges to State Modernization in 21st Century Europe' Estonian Institutional Grant [IUT 19-13] and the 'Web 2.0 and Governance: Institutional and Normative Changes and Challenges' Estonian Research Foundation grant [ETF 8571]. We dedicate this work to all those who are building the world they want, within the confines of the world they want to transcend.

DOI: 10.1057/9781137406897.0004

Part I
Theoretical Framework

1

Capitalism as a Creative Destruction System

Abstracts: *Many would argue that no other economic system than capitalism has produced so much wealth. On the other hand, some might claim that no other system has produced so much destruction. Others consider capitalism as a creative destruction system. This chapter discusses the theory of techno-economic paradigm shifts with the aim to recognize the dynamic nature of the capitalist system, and highlight the transition potential of new modes of social production and organization. Kostakis and Bauwens argue that the world is at a turning point where the excesses, the fallacies and the unsustainability of the current practices have to be recognized and appropriate regulatory changes have to be made, so that desperation and anger are turned into creation.*

Kostakis, Vasilis and Michel Bauwens. *Network Society and Future Scenarios for a Collaborative Economy*. Basingstoke: Palgrave Macmillan, 2014. DOI: 10.1057/9781137406897.0006.

DOI: 10.1057/9781137406897.0006

The capitalist mode of production has arguably created a political economy prone to crises. Following Harvey's (2012, p. 5) vivid narration, a typical day in the life of a capitalist begins with a certain amount of money and ends with a lot more. The next day, however, the capitalist has to think about how he is going to manage that surplus capital: will he reinvest the profits or will he spend them? As long as we are not speaking about monopolies (Baran and Sweezy, 1966), the fierce competition compels him to reinvest. If he does not, a competitor certainly will. Of course, a successful capitalist profits enough to maintain profitable expansion while also living a super-luxurious life. The constant search for new terrains of growth is a premise for the sustainability of the system. Capital accumulation must expand at a compound rate; according to Harvey (2012, p. 5), 'the result of perpetual reinvestment is the expansion of surplus production'. The capitalist faces a variety of problems during the aforementioned procedure. If wages were too high due to labor scarcity, for instance, fresh labor forces must be found or precarious living conditions must be artificially created, thus inducing a drop in wages, in order to keep the system in a growth trajectory. Furthermore, that new terrain of growth is enriched with the introduction of new means of production and technological and/or organizational innovations. New needs and wants are defined, distances between nation-states diminished, and the capitalist finds himself capable not only of discovering new natural resources but also of attracting new customers (Harvey, 2012, 2010; Perez, 2002). When purchasing power cannot serve an increasingly expanding economy, new credit-based financial instruments are invented. If the profit rate is low, sometimes companies merge, creating powerful conglomerates and, therefore, monopolies. If capital accumulation does not continue, then the system falls into a crisis: Capitalists are unable to find profitable paths of reinvestment; capital accumulation stagnates and its value decreases; massive unemployment, impoverishment and social turmoil are some of the potential consequences of a capitalist crisis.

But many would argue that no other economic system has produced so much wealth. On the other hand, some might claim that no other system has produced so much destruction. Others consider capitalism a creative destruction system. This book uses the theory of techno-economic paradigm shifts (TEPS) – gradually developed by Schumpeter (1982/1939, 1975/1942), Kondratieff (1979), Freeman (1974, 1996), and in particular Perez (1983, 1985, 1988, 2002, 2009a, 2009b) – as its point of departure to develop its narrative. This choice arguably helps to recognize the dynamic

DOI: 10.1057/9781137406897.0006

and changing nature of the capitalist system, in order to avoid any particular period of extrapolation as 'the end of history' in the fashion of Fukuyama (1992). Therefore, the aim is not to make capitalism crisis-free but to manage crises and soften blows. In other words, to form a successful 'creative destruction management' (Kalvet and Kattel, 2006), maximizing its creative power while minimizing its destructive force (Mulgan, 2013). One should be aware of many other theoretical alternatives, those of Marx for example, in understanding and acting within certain social, technological and economic processes. Interestingly, Marxist and neo-Schumpeterian theoretical approaches consider capitalism prone to crises, which are basic features of its normal functioning. However, the neo-Marxist critique (see Wolff, 2010; Harvey, 2007, 2010) puts emphasis on the inherent unsustainability of capitalism, aiming at a different system – 'modern society can do better than capitalism', Wolff (2010) postulates – whereas neo-Schumpeterians, such as Perez (2002) or Freeman (1974; 1996), see crises as a chance to move the capitalist economy forward. This book is an integrative attempt at highlighting the potential of new modes of social production and organization immanent in capitalism but which, in the long term, might transcend the dominant system.

If we follow Schmoller (1898/1893), the main figure of the German Historical School, history is the laboratory of the economist. Despite the unquestionable uniqueness of each historical period in socio-economic development, the theory of TEPS accepts recurrence as a frame of reference and, having each period's uniqueness as the object of study, tries to interpret the potential and the direction of change (Perez, 2002). Moreover, it embraces the Schumpeterian (1982/1939) understanding of economy as 'an interdependent sequence of dynamic forces of change and static equilibrating forces' (Drechsler et al., 2006, p. 15). The essential fact about capitalism is the process of creative destruction incessantly revolutionizing the economic structure from within, destroying the old one while creating a new one (Schumpeter, 1975/1942). Each techno-economic paradigm (TEP) is based on a constellation of innovations, both technical and organizational, which are the driving force behind economic development (Perez, 1983). Each TEP plays the central role in a recurring pattern of cyclical movement: from gilded ages to golden ages; from an initial installation period, through a collapse and recession that signify the turning point, to a full deployment period (Perez, 2002, 2009a). Therefore, in the Perezian framework (2002, 2009a), progress in capitalism takes place by going through various successive

DOI: 10.1057/9781137406897.0006

great surges of development which are driven by successive technological revolutions. Each of these overlapping great surges of development, lasting approximately 40–60 years, is the process by which a technological revolution and its paradigm propagate across the economy, 'leading to structural changes in production, distribution, communication and consumption as well as to profound and qualitative changes in society' (Perez, 2002, p. 15).

According to the TEPS theory, the world has experienced five technological revolutions during the past three centuries: the first industrial revolution based on machines, factories and canals (initiated in 1771; birthplace: Britain); the age of steam, coal, iron and railways (1829; Britain); the age of steel and heavy engineering (1875; Britain, USA, and Germany); the age of automobile, oil, petrochemicals and mass production (1908; USA); and the age of information technology and communication (1971; USA). Each of these processes evolved 'from small beginnings in restricted sectors and geographic regions', and ended up 'encompassing the bulk of activities in the core country or countries and diffusing out towards further and further peripheries, depending on the capacity of the transport and communications infrastructures' (Perez, 2002, p. 15).

A great surge of development consists of four phases, which, although not strictly separated, can be identified as sharing common characteristics throughout history (Figure 1.1).

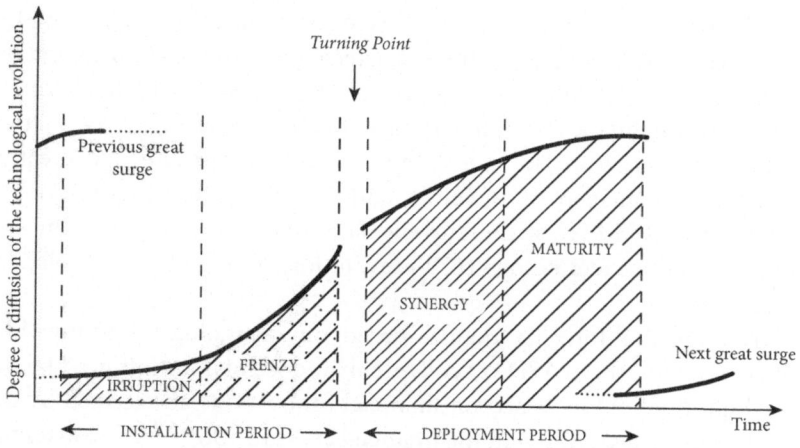

FIGURE 1.1 *Recurring phases of each great surge in the core countries*

Source: Based on Perez, C. (2002) *Technological Revolutions and Financial Capital: The Dynamics of Bubbles and Golden Ages* (Cheltenham: Edward Elgar Pub), p. 48.

DOI: 10.1057/9781137406897.0006

First, we have irruption (technological explosion), or the initial development of new technologies in a world where the bulk of the economy is made of old, maturing and declining industries. Frenzy follows, which is the rapid development of technology requiring a great deal of finance (this is when financial bubbles are created). These two first phases constitute the installation period of the new TEP, when finance and greed prevail and the paper economy decouples from the real one. Next, turbulent times arrive – that is, collapse, recession and instability. This is what Perez calls the turning point: neither a phase nor an event, but rather a process of contextual shift, where institutional changes for the deployment period of the newly installed paradigm take place. Institutional innovations occur, which enable economies to take advantage of new technology across all sectors, and in turn to spread the benefits of this new wealth-creating potential widely across society. These synergies appear in the early stages of deployment (synergy phase) until they approach a ceiling (maturity phase) in productivity, new products and markets. Once that ceiling is hit, social unrest and confrontations will occur while conditions for the installation of the new paradigm, based on the next technological revolution, are set.

Perez (2009b) highlights the special nature of major technological bubbles (MTB), which are endogenous to the process by which society and the economy assimilate each great surge. The MTB tend to take place along the diffusion path of each technological revolution: from the installation period, when the new constellation of technologies is tested and investment is defined by the short-term goals of financial capital (so a rift between real values and paper values occurs), to the deployment period, when financial capital is brought back to reality, production capital takes the lead and the state is called to make effective 'creative destruction management' (Kalvet and Kattel, 2006). Perez (2009b) argues that the MTB of the current TEP, that is the information and communications technology (ICT) revolution, occurred in two episodes (Figure 1.2).

First was the Internet mania, based on technological innovation, which ended in the NASDAQ collapse in 2000. This was followed by the easy liquidity bubble, based on financial innovations accelerated by the new technologies, ending in the financial crisis in 2007–08. The essential implication of Perez' (2009b, p. 803) argumentation is that 'what we are facing is not just a financial crisis but rather the end of a period and the need for a structural shift in social and economic context to allow for

DOI: 10.1057/9781137406897.0006

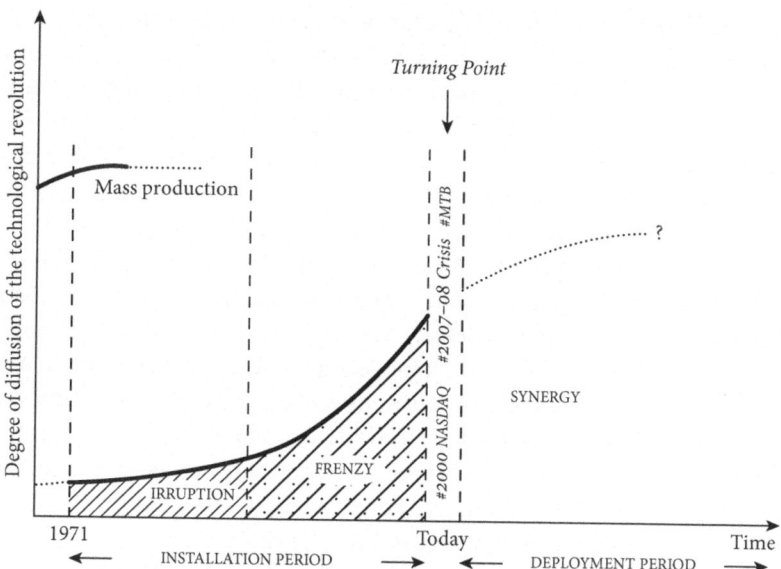

FIGURE 1.2   *The current ICT-driven techno-economic paradigm: The major technological bubbles at the turning point and a deployment period to come*

Source: Based on Perez, C. (2002) *Technological Revolutions and Financial Capital: The Dynamics of Bubbles and Golden Ages* (Cheltenham: Edward Elgar Pub), p. 48.

continued growth under this paradigm'. Moreover, Perez' (2009b) essay on the double bubble, aligned with the TEPS theory, is used as a point of departure that treats the current situation as not just another passing recession, and sets the ground for tentative proposals concerning the second half of the ICT revolution's wealth-generating potential.

Since the introduction of the microprocessor (California, November, 1971), and after a nearly 30-year-long paroxystic culmination of market experimentation and moments of Galbraithian (1993) irrationality, we find ourselves in the aftermath of two major bubbles and, arguably, in the midst of a major capitalist crisis (Fuchs et al., 2010, p. 193). In other words, we are witnessing, as we will later see, the swing of the pendulum from extreme individualism to collective, synergistic well-being. The whole system is trying to recompose (Perez, 2002), while political unrest (e.g., the EU coherency crisis triggered by the debt crisis) and protests (from the Indignados movement in Spain and the protest movement in Greece to the Occupy Wall Street movement in the USA) are erupting

DOI: 10.1057/9781137406897.0006

globally. However, this book's goal is neither to describe the strands and ramifications of the current crisis, as this has been done elsewhere (see Harvey, 2007, 2010; Chomsky, 2011; Funnell, Jupe and Andrew, 2009; Stiglitz, 2010), nor to indicate historical parallels in previous turning points within capitalism, as Perez has done that in detail in her 2002 book. It can be claimed, though, that the two bubbles at the turn of this century recall the 1929 depression in that they share one fundamental characteristic: the structural tensions within capitalism make the system, at least in its current form, unsustainable. The world is arguably at a crossroads where the excesses, the fallacies and the unsustainability of the current practices need to be recognized; appropriate regulatory changes have to be made where the usual recipes for confronting tensions fail; and conditions where production capital is put in control, greater social cohesion is achieved, and desperation and anger turn into creation must be facilitated (Perez, 2002, 2009a, 2009b). In other words, this turning point is a time of indeterminate realization of the full potential of the current ICT-driven paradigm, creating the new fabric of the economy and overcoming the tensions that caused this premature saturation (Perez, 2002).

DOI: 10.1057/9781137406897.0006

2
Beyond the End of History: Three Competing Value Models

Abstract: *At the current turning point of the ICT-based techno-economic paradigm and within the present political economy, this chapter argues, there are three different value models competing for dominance, which influence the way that the institutional recompositions will take place. One form is still dominant, but rapidly declining in importance; a second form is reaching dominance; and a third is emerging. This chapter discusses the decline of the first competing value model, that of the classic capitalist economy based on labor value and proprietary forms of knowledge.*

Kostakis, Vasilis and Michel Bauwens. *Network Society and Future Scenarios for a Collaborative Economy.* Basingstoke: Palgrave Macmillan, 2014.
DOI: 10.1057/9781137406897.0007.

Have we already lived through the end of history with the fall of the Berlin wall in 1989–90? Is the capitalist mode of production in the final stage of human progress? Or are we currently living in the end times with capitalism approaching its terminal crisis? According to Žižek (2010, p. x), the dominant system is unable to face its internal imbalances and its failures: the ongoing ecological crisis as well as the emergence of new forms of apartheid, walls and slums. Capitalism transforms not because of its failures but because of its successes, neo-Schumpeterians might reply, and now it is high time we created virtuous circles of production that would allow the system to reinvent itself once again. The environmental crisis can be seen as an opportunity for investment and sustainable growth (Gore, 2013). In the meantime, a new type of capitalism, named 'cognitive capitalism', arises in which 'the object of accumulation consists mainly of knowledge' that is now the basic source of value (Boutang, 2012, p. 57). The industrial mode of production is becoming obsolete, and the 'network' is the main pattern of organizing production and socio-political relations (see Castells, 2000, 2003, 2009). Peer-to-peer (P2P) technologies and renewable energy merge, creating an energy Internet and, thus, inaugurating a third industrial revolution (Rifkin, 2011). On top of that, one may add another disruptive technological cluster, the 'Internet of Things', which could help 'humanity reintegrate itself into the complex choreography of the biosphere, and by doing so, dramatically increases productivity without compromising the ecological relationships that govern the planet' (Rifkin, 2014, p. 13). Others (see Anderson, 2012) point to emerging desktop manufacturing technologies, such as the three dimensional (3D) printing, and consider them the pervasive technological cluster which will trigger a new industrial revolution. Success in taking advantage of these transformations, and, at least in theory, the benefits of new wealth creating potential will spread more widely across society.

This book argues that, at the current turning point of the ICT-based TEP and within the present political economy, there are three different value models competing for dominance, which influence the way institutional recompositions will take place. One form is still dominant, but rapidly declining in importance; a second form is reaching dominance, but carries within itself the seeds of its own destruction; and a third is emerging, but needs vital new policies in order to become dominant (Figure 2.1).

DOI: 10.1057/9781137406897.0007

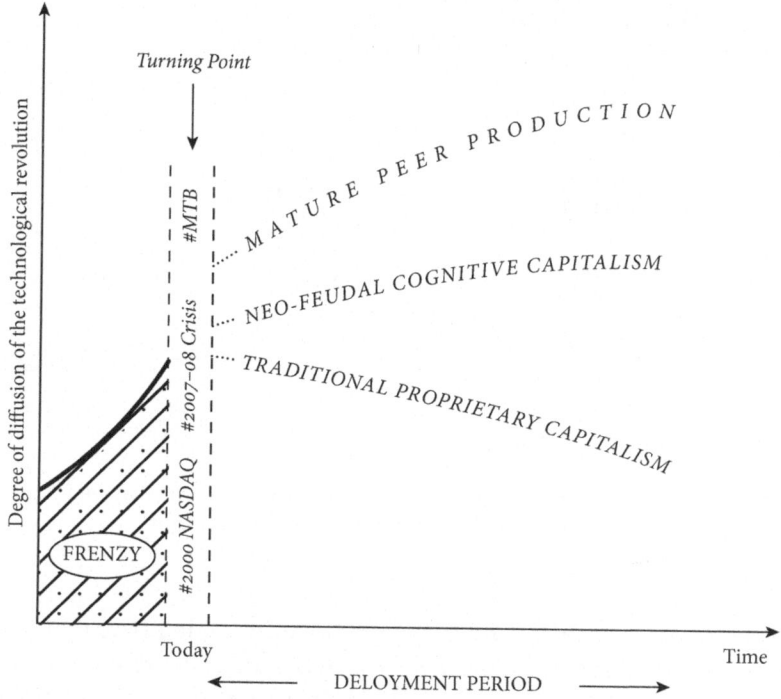

FIGURE 2.1 *Three competing value models*

The first is the classic capitalist economy based on labor value and proprietary forms of knowledge, which dominated the industrial phase of capitalism. The value model of the traditional proprietary capitalism is based on the premise that workers create value in their private capacity as providers of labor (Figure 2.2). This value is captured and realized in the market by capital, which dominates the extraction of surplus value. In the old neoliberal vision, the state becomes a market state which protects the privileged interests of property owners; and civil society is a 'rest category' – a sphere of minor importance as is evidenced in the use of our language (nonprofits, nongovernmental). The de-skilling of workers – what was once artisanal production knowledge but which is now codified in the production process itself – characterizes this form. Labor becomes an appendage to the ecosystem of machines. In this division between labor and capital, managerial and engineering layers handle collective production on behalf of the owners of capital. At first,

DOI: 10.1057/9781137406897.0007

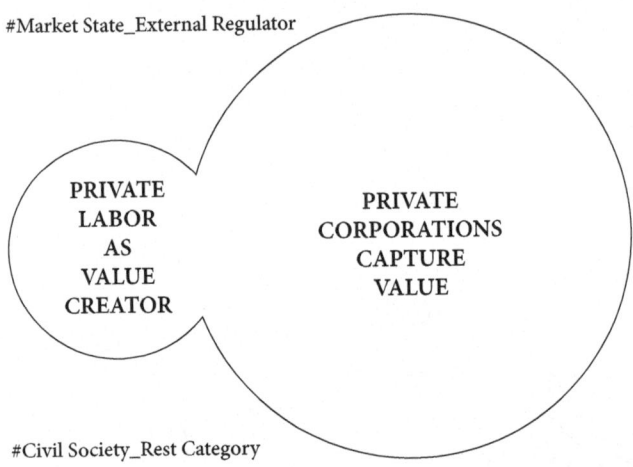

PRIVATE
LABOR
AS
VALUE
CREATOR

PRIVATE
CORPORATIONS
CAPTURE
VALUE

#Market State_External Regulator

#Civil Society_Rest Category

FIGURE 2.2   *The value model of the traditional proprietary capitalism which dominated the first phase of the current techno-economic paradigm*

this is largely industrial capital, though financial capital rises rapidly to prominence. Codified knowledge is proprietary and value is increasingly captured as intellectual property (IP) rent. However, industrial profit, based on the direct extraction of surplus value, is the dominant form of value capture, and there is partial redistribution in the form of wages.

Often, once a social (labor) movement takes form and becomes powerful and influential, the state redistributes taxable wealth to the workers as consumers and citizens in the form of social provisions (pensions, unemployment benefits, health care reimbursements etc.). This happened extensively during 1945–80, manifested by the rise of the welfare state and Keynesian policies, especially in the Western world. Since 1980, under contemporary conditions of labor weakness in the de-industrializing developed countries, the state has been redistributing wealth to the financial sector and creating conditions of debt dependence for the majority of the population. In this neoliberal format, which became dominant after 1980 before the emergence of civic peer networks on the eve of the 21st century (Benkler, 2006; Bauwens, 2005), the part of labor became stagnant and most of the value was streamed toward financial capital. The credit system developed into an increasingly important means to maintain the fictitious buying power of consumers

DOI: 10.1057/9781137406897.0007

and, therefore, the primary means of surplus realization through debt dependency and servicing.

We argue that this value model of traditional proprietary capitalism, dominant in the installation period of the current TEP, is approaching its terminal point. Its inherent unsustainability is manifested in a twofold problem. On the one hand, industrial capitalism considers nature to be a perpetually abundant resource; that is, it is based on a false notion of material abundance in a finite world. On the other hand, the traditional, industrial version of cognitive capitalism enforces the idea that intellectual, scientific and technical exchange should be subject to strong proprietary constraints. In that way, an artificial scarcity of knowledge is created, subjecting innovation to legal restrictions and allowing for profit maximization and, hence, capital accumulation. Thus appears the paradoxical but also dramatic contradiction of the present, dominant system: while it is rapidly overburdening the carrying capacity of the planet, it simultaneously inhibits the solutions humanity might find for it. For example, the dramatic increase in patents has not been paralleled by an increase in technological innovation: 'there is no empirical evidence that they [patents] serve to increase innovation and productivity, unless productivity [or innovation] is identified with the number of patents awarded' (Boldrin and Levine, 2013, p. 3). 'In the long run', Boldrin and Levine (2013, p. 7) argue, 'patents reduce the incentives for current innovation because current innovators are subject to constant legal action and licensing demands from earlier patent holders.' The process of innovation relies upon building on former innovations. Therefore, the broader the pool of accessible ideas, the more chances there are for innovation (Brynjolfsson and McAfee, 2011). To recap, this combination of quasi-abundance and quasi-scarcity destroys the biosphere and hampers the expansion of social innovation and a 'free culture' (as described in Lessig, 2004), and this situation, arguably, must be reversed.

The recent crises have brought scholars from various traditions and schools to agree that the global economy is currently at a turning point within the ICT-driven TEP. In this book, we deal with the remaining two competing value models. These are more synchronized with the main characteristics of the current TEP, and they seem to introduce less-fragile alternative approaches for development in the deployment period. The second form is the neo-feudal cognitive capitalism, in which proprietary forms of knowledge are in the process of being displaced by emerging

DOI: 10.1057/9781137406897.0007

forms of peer production (Benkler, 2006; Bauwens, 2005), but under the dominance of financial capital. We will describe how this process is well under way. The third is the hypothetical form of mature peer production under civic dominance, whose stems are already emerging through the interstices of the dominant system.

DOI: 10.1057/9781137406897.0007

3
The P2P Infrastructures: Two Axes and Four Quadrants

Abstract: *The P2P infrastructures, such as the Internet, are those infrastructures for communication, cooperation and common value creation that allow for permission-less interlinking of human cooperators and their technological aids. It has been assumed that such infrastructures are becoming the general conditions of work, life and society. In this context, this chapter introduces a four-scenario approach which attempts to simplify possible outcomes by using two axes or polarities (global versus local orientation; centralized versus distributed control of the infrastructure). Each quadrant stands for a certain scenario where each technological regime (namely, netarchical capitalism, distributed capitalism, resilient communities, and global Commons) is dominant.*

Kostakis, Vasilis and Michel Bauwens. *Network Society and Future Scenarios for a Collaborative Economy.* Basingstoke: Palgrave Macmillan, 2014. DOI: 10.1057/9781137406897.0008.

The P2P infrastructures, such as the Internet, are those infrastructures for communication, cooperation and common value creation that allow for permission-less interlinking of human cooperators and their technological aids. We argue that such infrastructures are becoming the general conditions of work, life and society (see Bauwens, 2005). Of course, one should be aware of the danger of 'Internet-centrism' (Morozov, 2012) and the perception that the Internet is the solution to all of humanity's problems. However, change is unlikely to occur without sufficient ICT penetration since, as has become evident, various aspects of complex human nature can be amplified and telescoped by the Internet (MacKinnon, 2012). P2P relational dynamics, which sometimes seem to epitomize the old slogan 'Jeder nach seinen Fähigkeiten, jedem nach seinen Bedürfnissen!' [from each according to his ability, to each according to his need], are based on the distribution of the productive forces. First, the means of information, immaterial production, that is the networked computers, and now the means of physical manufacturing, that is, machines that produce physical objects, are being distributed and interconnected. Just as networked computers democratized the means of production of information and communication, the emergent elements of networked micro-factories or what some (see Kostakis, Fountouklis and Drechsler, 2013; Anderson, 2012; Rifkin, 2014) call desktop manufacturing, such as 3D printing and computer-numerical-control (CNC) machines, are democratizing the means of making.

Of course, this process is not without its problems. In a time of extreme polarization and with no equilibrium reached in regard to global governance of the Internet (Mueller, 2010), we have witnessed conflicts over the control and ownership of distributed infrastructures. For example, the Internet, the world's largest ungoverned space (Schmidt and Cohen, 2013), has become a highly contested political space (MacKinnon, 2012). On the one side, peer production signals for some fundamental changes to take place juxtaposing them against an old order that should be cast off (Bauwens, 2005; Benkler, 2006). On the other, the proposed legislations of ACTA/SOPA/PIPA that enforce strict copyright; the attempts at surveillance, public opinion manipulation, censorship and the marginalization of opposite voices by both authoritarian and liberal countries (MacKinnon, 2012); and 'the growing tendency to link the Internet's security problems to the very properties that made it innovative and revolutionary in the first

DOI: 10.1057/9781137406897.0008

place' (Mueller, 2010, p. 160) are only some of the reasons that have made some scholars (see Zittrain, 2008; MacKinnon, 2012) worry that digital systems may be pushed back to the model of locked-down devices or centrally controlled information appliances. Hence, there appears to be a battle emerging among agents (several governments and corporations), which are trying to turn the Internet into a tightly controlled information medium, and user communities who are trying to keep the medium independent.

This book attempts to simplify possible outcomes by using two axes, or polarities, which give rise to four possible scenarios. Each quadrant stands for a certain scenario where each technological regime is dominant. This does not exclude the presence of the rest; however, the dominant regime defines the kind of political economy which may prevail. Value regimes are more or less associated with technology regimes, since the forces at play want to protect their interests through the control of technological and media platforms, which encourage certain behaviors and logics, but discourage others. The powers over technological protocols and value-driven design decisions are used to create technological platforms that match proprietary interests. Even as P2P technologies and networks are becoming ubiquitous, ostensibly similar P2P technologies have very different characteristics which lead to different models of value creation and distribution, and thus to different social and technological behaviors. In networks, human behavior can be subtly – or not so subtly – influenced by design decisions and invisible protocols created in the interest of the owners or managers of the platforms.

Figure 3.1 is organized around two axes, which determine at least four distinct possibilities. The first top-down axis distinguishes centralized technological control (and an orientation toward globality) from distributed technological control (and an orientation toward localization); the horizontal axis distinguishes a for-profit orientation (where any social good is subsumed to the goal of shareholder profit), from for-benefit orientations (where eventual profits are subsumed to the social goal).

The four-scenario approach has been widely used as an exploratory tool that allows for fruitful discussions on policymaking (van der Heijden, 2005; Leigh, 2003) and sustainable strategic planning and development (Godet, 2000; Kelly, Sirr and Ratcliffe, 2004). Each scenario has a descriptive role and outlines tentative political economies with the aim of sparking the imagination and serving as a route map for the future

DOI: 10.1057/9781137406897.0008

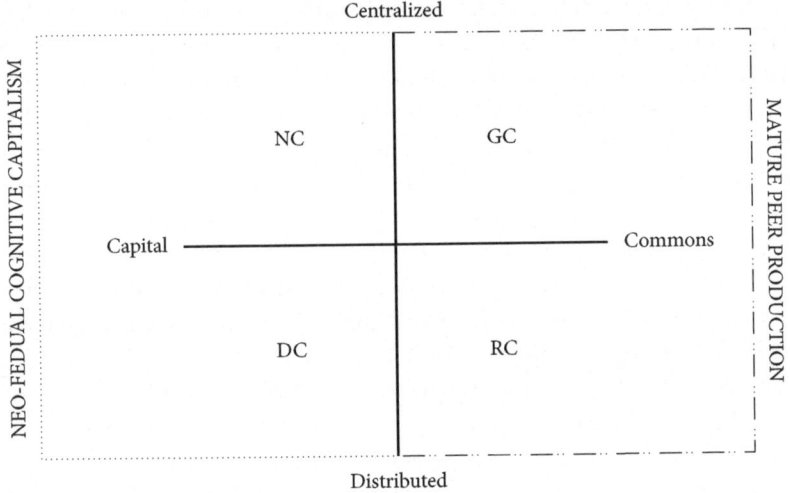

FIGURE 3.1   *Two axes and four future scenarios*

(Miles, 2004). Using scenarios is like rehearsing the future, according to Schwartz (1996). By rehearsing these future scenarios, organizations, states and the civil society can adapt to what is happening and anticipate and influence what could transpire. In accordance with van der Heijden et al. (2002) and Schwartz (1996), our scenario framework consists of two dimensions which have high levels of uncertainty and are crucial to future developments. The first axis presents the polarity of centralized versus distributed control of the productive infrastructure, whereas the second axis relates an orientation toward the accumulation of capital versus an orientation toward the accumulation or circulation of the Commons.

Within this context the following four future scenarios for economy and society are introduced: netarchical capitalism (NC), distributed capitalism (DC), resilient communities (RC) and global Commons (GC). Netarchical and distributed capitalism differ in the control of the productive infrastructure but both are oriented toward capital accumulation and, thus, are parts of the wider value mode of cognitive capitalism. They actually form the mixed model of neo-feudal cognitive capitalism. On the other, resilient communities and the global Commons reside in the, one might say auspicious, hypothetical model of mature peer production under civic dominance (right quadrants). The next parts shed

DOI: 10.1057/9781137406897.0008

light on each scenario in separate chapters, also discussing the coexistence of each pair of models sharing a common orientation. Moreover, Part III attempts to introduce a few preliminary general principles for policymaking, and put forward some general policy recommendations with the goal of moving from the left side of the quadrants to the right. Or, to put it in the terms of the TEPS theory, to realize the full potential of an ICT-driven TEP while maximizing the benefits from technological progress for the largest part of society.

DOI: 10.1057/9781137406897.0008

Part II
Cognitive Capitalism

Cognitive capitalism refers to the process by which information (data, knowledge, design or culture) is privatized and then commodified as a means of generating profit for capital. In this new phase of capitalism, traditional processes of material production and distribution are overtaken by the control of information as the driving force of capital accumulation (see Boutang, 2012; Bell, 1973; Drucker, 1969; for a critical analysis, see Webster, 2006). Of course, we should be aware of Federici and Caffentzis' (2007, p. 70) remark that notions such as 'cognitive labor' and 'cognitive capitalism' represent 'a part, though a leading one, of capitalist development and that different forms of knowledge and cognitive work exist that cannot be flattened under one label'. In general, one could argue that capitalism, in the past, was primarily concerned with the commodification of material. Essential to this process was the gradual enclosure and privatization of the material Commons, including pasture lands, forests and waterways that had been used in common since time immemorial (for an analysis of the 1700–1820 enclosure in England, see Neeson, 1993). In our time, capitalism entails the enclosure and commodification of the immaterial: knowledge, culture, DNA, airwaves, even ideas (for an account of the 'second enclosure movement', see Boyle, 2003b). Ultimately, the driving force of capitalism in our age is the eradication

DOI: 10.1057/9781137406897.0009

of all Commons and the commodification of all things. The colonization and appropriation of the public domain by capital is arguably at the heart of the new enclosures. This process is sustained and extended through the complex and ever-evolving web of patents, copyright laws, trade agreements, think tanks, and government and academic institutions that provide the legal, policy and ideological frameworks that justify all this (for a critical perspective on strict intellectual property see Lessig, 2004; Boldrin and Levine, 2013; Patry, 2009; Bessen and Meuer, 2009). Above all, the logic of this process is embedded in the values, organization and operation of the traditional capitalist firm.

In the new vision of cognitive capitalism, which represents this book's second competing value model, networked social cooperation consists of mostly unpaid activities that can be captured and financialized by proprietary 'network' platforms. Social media platforms almost exclusively capture the value of their members' social exchange, and distributed labor, such as crowdsourcing, tends to reduce the average income of the producers (for an overview of crowdsourcing's labor markets, digital labor and the dark side of the Internet in general, see the collective book edited by Scholz, 2012). The 'netarchical' (meaning, the hierarchies within the network which own and control participatory platforms) version of networked production, here, creates a permanent precariat and reinforces the neoliberal trends. Projects such as the P2P currency Bitcoin and the Kickstarter crowdfunding platform are representative examples of more distributed developments which embrace the idea that 'everyone can become an independent capitalist'. Under this model, P2P infrastructures are designed to allow the autonomy and participation of many players, but the main focus is still profit maximization. Next we deal with the two forms of the neo-feudal model of cognitive capitalism (left quadrants), which are based on various technological regimes dependent on the structure of every project's back-end. User-oriented technological systems generally have two sides. The front-end is the side that users interact with, and is the only side visible to them. In other words, it is the interface with the other users and with the system itself. The back-end, however, is the technological underpinning that makes it all possible. This is engineered by the platform owners and is invisible to the user. Hence, a front-end which enables a P2P social logic among users can often be highly centralized, controlled, and proprietary on the back-end; forming an invisible techno-social system that profoundly influences the behavior of those using the front-end, by setting limits on

DOI: 10.1057/9781137406897.0009

what is possible in terms of human freedom. As we will see in Chapters 4 and 5, a truly free P2P logic at the front-end is highly improbable if the back-end is under exclusive control and ownership. This part concludes with Chapter 6, where the potentialities of this value model are discussed.

DOI: 10.1057/9781137406897.0009

4
Netarchical Capitalism

Abstract: *This chapter describes the first technological regime/future scenario which develops within the context of a new-feudal form of cognitive capitalism. 'Netarchical capitalism' matches centralized control of a distributed infrastructure with an orientation toward the accumulation of capital. For Kostakis and Bauwens, the netarchical capital is that fraction of capital which enables cooperation, but through proprietary platforms that are under central control. While individuals share through these platforms, they have no control over the design and the protocol of these networks/platforms, which are proprietary. Typically under conditions of netarchical capitalism, while sharers directly create or share use value, the monetized exchange value is realized by the owners of capital. This arguably creates a longer-term 'value crisis', since the value creators are not rewarded.*

Kostakis, Vasilis and Michel Bauwens. *Network Society and Future Scenarios for a Collaborative Economy*. Basingstoke: Palgrave Macmillan, 2014. DOI: 10.1057/9781137406897.0010.

The period since the 1990s saw the birth of a mixed regime. Civic inter-networks (systems of interconnected networks) became increasingly available to a wider population, and other forms of networked value creation became possible. Use value has been created independently of the private industrial and financial system, through different forms of peer production and networked value creation. This creative process has taken place in the form of civic contributions, where immaterial use value is deposited in common pools of knowledge, code and design. In 'pure' peer production, this immaterial value is contributed and deposited into common pools by voluntary or paid contributors. The for-benefit associations, such as the Free/Libre/Open Source Software (FLOSS) foundations, enable continued cooperation; and entrepreneurial coalitions of mostly for-profit capitalist enterprise capture the added value in the marketplace. For example, the cases of the International Business Machines corporation (IBM) and Linux is well-known and widely discussed (see Tapscott and Williams, 2006; Coleman and Hill, 2004; IBM, 2010). This coalition shows how a firm entered the FLOSS ecology and invested monetary and human capital (improving the reliability of Linux by testing code, error handling etc.) in the development of FLOSS. IBM, according to its corporate report (2010), holds significant roles in a large number of FLOSS projects such as in the development of the Linux Kernel, Apache, Eclipse or Ubuntu, working closely with Red Hat, a leading distributor of the Linux enterprise. On the one hand, IBM's involvement enhanced the quality of the outputs and the sustainability of the projects, creating chances for wage labor for some of the most active and skillful Linux developers in the market economy. On the other, the rewards from such an involvement have been considerable for IBM. According to Tapscott and Williams (at least at the time of their writing in 2006) the firm would spend about $100 million per year on general Linux development. So if the Linux community produces use value of $1 billion (if it were to be produced by paid labor), and even half of that is useful to IBM, then the firm gains $500 million of software development for an investment of $100 million (Tapscott and Williams, 2006). 'Linux gives us a viable platform uniquely tailored to our needs for twenty percent of the cost of a proprietary OS' says Cawley, IBM's business development executive at that time, in Tapscott and Williams (2006, p. 81). To put the matter bluntly, IBM would pay $2 to ten employees but would get a value of more than $20 by many more than ten contributors, from whom a considerable number would participate on a voluntary basis. In this model,

DOI: 10.1057/9781137406897.0010

there is a continued creation of use value in the public sphere and, thus, an accumulation or a circulation of the Commons based on open input, participatory processes of production and Commons-oriented output. However, the accumulation of capital still continues through the form of labor and capital in the entrepreneurial coalitions. It becomes obvious that an increasing amount of voluntary labor is extracted in this process.

In the so-called sharing economies of networked value characterized by networking processes which take place over proprietary platforms, the use value is created by the social media users, but their attention is what creates a marketplace where that use value becomes extracted exchange value. In the realm of exchange value, this new form of netarchical capitalism may be interpreted as hyper-exploitation, since the use value creators go totally unrewarded in terms of exchange value, which is solely realized by the proprietary platforms. For instance, Facebook and Google, perhaps the two bigger netarchical capitalists, abandon direct production and instead create and maintain platforms which allow people to produce. They rely much more marginally on IP protection, but rather allow P2P communication while controlling its potential monetization through their ownership of the platforms for such communication. Typically, the front-end is P2P, in that it allows P2P sociality, but the back-end is controlled. The design is in the hands of the owners, as are the private data of the users, and it is the attention of the user-base that is marketed through advertising. The financialization of cooperation is still the name of the game. The back-end of these platforms, which serve as attention pools, is generally a centralized system where personal data is privatized. The monetization of the surplus value produced is exclusionary, keeping the users/producers out of that process. Almost everything is controlled by the owners of the platforms and there is an unequal distribution of power among owners and users. The same applies in other proprietary platforms, such as Airbnb, a platform that helps people to rent out lodging, including private rooms, entire apartments, boats, tree houses, private islands and other properties. In other words, it commodifies things, that is, idle resources, that were not previously commodified. If one looks carefully at the back-end of Airbnb's productive structure, he/she would realize that there is neither collaborative production nor governance, and the control rests with the owners of the platform. In essence, platform owners, who are crucially dependent on the trust of user communities, exploit the aggregated attention and input

DOI: 10.1057/9781137406897.0010

of the networks in different ways, even as they enable it. In addition, such platforms are dangerous as trustees of any common value that might be created, due to their speculative nature and the opaque architecture (closed source) of their platforms (Kostakis, 2012). The parasitic nature of this mode becomes evident by the fact that an empty networking platform is arguably a valueless platform. In addition to this, search engines and social networks limit the diversity of information sources so as to please their advertising customers, potentially minimizing the development of critically thinking citizens (Pariser, 2011). To recap, we call 'aggregated distribution' the productive models which are followed by corporations such as Google, Facebook, Airbnb or even IBM. Of course, it is important to emphasize that each netarchical project has its own special characteristics and peculiarities and it is difficult, if not impossible, to provide an all-inclusive description. However, what these projects have in common is that while their front-ends (whether the platform's infrastructure, see Facebook or Airbnb, or a P2P practice that the company may follow, see IBM) might be distributed, they are based on certain technological regimes of centralized back-ends while having a for-profit orientation with exclusionary financialization (Figure 4.1).

Further, in the form of crowdsourced marketplaces, capital abandons the labor form and externalizes risk onto the freelancers. Crowdsourcing

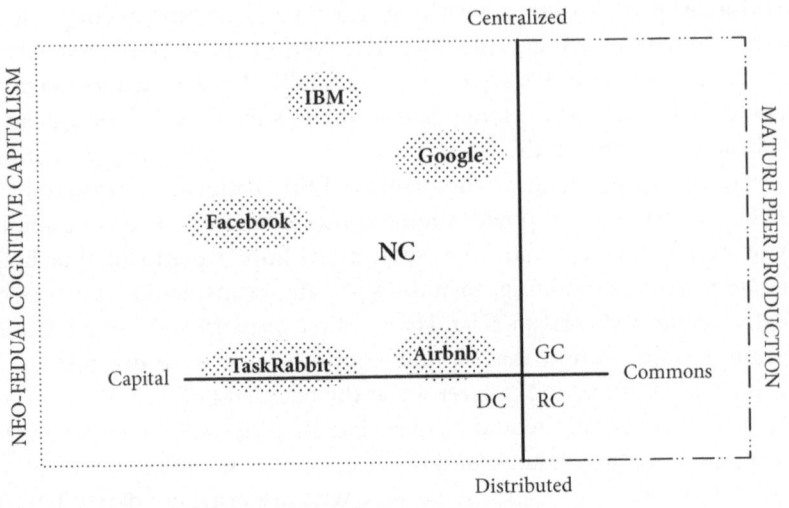

FIGURE 4.1   *The netarchical capitalism quadrant*

DOI: 10.1057/9781137406897.0010

economies are not very different to the sharing ones in that users still 'share' information, in a way. Compared with the sharing/aggregation economies, the profit motive for users is a bit stronger here, mainly in the form of a prize (Kostakis, 2012). Howe (2008) offers case histories such as iStock, a community-driven source for stock photography, and InnoCentive, where firms offer cash prizes for solving some of their thorniest development problems. Other crowdsourcing platforms include 99designs and DesignCrowd, which both deal with design (from logo design to T-shirt design). We consider crowdsourcing projects as 'disaggregated distribution', because the workers are isolated freelancers competing without collective shared IP. For instance, think of a crowd-sourced logo production: the crowdsourcing company will choose the best logo out of, say, 50 logos, and the remaining 49 will often be trashed. No production of common, shared value takes place. Another typical example could be the 'skills' marketplace TaskRabbit, where workers cannot communicate with each other, but clients can. The producers are isolated as there is no connection between the supply side and the demand side. The project platform is designed to favor demand, while the network is controlled by the owners of the platform.

Under this regime of cognitive capitalism, which includes both aggregated and disaggregated distribution, use value creation expands exponentially but exchange value only rises linearly and is almost exclusively realized by capital, giving rise to forms of hyper-exploitation. We could call this value regime neo-feudal, because it often relies on unpaid 'corvée' (i.e., statute labor) and creates wide-spread debt peonage. Ownership is replaced by access, diminishing the sovereignty that comes with property, and creating dependencies through the one-sided licensing agreements in the digital sphere. We would argue that it creates a form of hyper-neoliberalism. While in classic neoliberalism labor income stagnates, in hyper-neoliberalism society is deproletarized, that is, wage labor is increasingly replaced by isolated and mostly precarious freelancers; more use value escapes the labor form altogether. Under the mixed regime of cognitive capi-talism in its netarchical form, networked value production grows, and has many emancipatory effects in the social field of use value creation. However, this is in contradiction with the field of exchange value realization, where hyper-exploitation occurs. In other words, there is an increased contradiction between the proto-mode of production, which is peer production, and associated forms of networked value

DOI: 10.1057/9781137406897.0010

creation with the relations of production, which remain under the domination of financial capital.

To sum up, we define 'netarchical capitalism' as the first combination (upper-left) which matches centralized control of a distributed infrastructure with an orientation toward the accumulation of capital. Netarchical capital is that fraction of capital which enables and empowers cooperation and P2P dynamics, but through proprietary platforms that are under central control. While individuals will share through these platforms, they have no control, governance or ownership over the design and the protocol of these networks/platforms, which are proprietary. Typically, under conditions of netarchical capitalism, sharers will directly create or share use value while the monetized exchange value will be realized by the owners of capital. Whereas in the short term it is in the interest of shareholders or owners, this also creates a longer-term value crisis for capital, since the value creators are not rewarded (or if they are, not in a decent way). They no longer have the purchasing power to acquire the goods that are necessary for the functioning of the physical economy.

On the one hand, in this technological regime a sector of capital has, to some significant degree, liberated itself of the need for proprietary forms of knowledge, but on the other, it has actually increased the level of surplus value extraction. At the same time, use value escapes more and more from its dependency on capital. This form of hyper-neoliberalism creates a crisis of value. The emergence of P2P models of production, based on the non-rivalrous nature and low marginal cost of digital information reproduction, coupled with the increasing unenforceability of IP laws, means that capital is incapable of realizing returns on ownership in the cognitive realm. In short, the creation of non-monetary value is exponential, whereas the monetization of such value is linear. There is a growing discrepancy between the direct creation of use value through social relationships and collective intelligence, but only a fraction of that value can actually be captured by business and money. Innovation is becoming social and diffuse; an emergent property of networks rather than an internal R&D affair within corporations. Hence, capital is becoming an a posteriori intervention in the realization of innovation rather than a condition for its occurrence, while more and more positive externalizations are created from the social field. What this announces is not only a crisis of value, most of which is 'beyond measure', but also essentially a crisis of accumulation of capital. Furthermore, we lack

DOI: 10.1057/9781137406897.0010

a mechanism for the existing institutional world to re-fund what it receives from the social world. On top of all of that, we have a crisis of social reproduction: peer production is collectively sustainable, but not individually (for an in-depth examination of these correlated issues, see Arvidsson and Pietersen, 2013).

DOI: 10.1057/9781137406897.0010

5
Distributed Capitalism

Abstract: *This chapter discusses the second technological regime/future scenario which develops within the context of a new-feudal form of cognitive capitalism. For Kostakis and Bauwens, 'distributed capitalism' matches distributed control over the infrastructure with a focus on capital accumulation. Under this technological regime, P2P infrastructures are designed in such a way as to allow the autonomy and participation of many players. Any Commons is a by-product or an afterthought of the system, and personal motivations are driven by exchange, trade and profit. Various P2P developments can be seen within this context, striving for a more inclusionary distributed and participative capitalism. Though they can be considered as part of an anti-systemic entrepreneurialism directed against the monopolies and predatory intermediaries, they retain the focus on profit making.*

Kostakis, Vasilis and Michel Bauwens. *Network Society and Future Scenarios for a Collaborative Economy.* Basingstoke: Palgrave Macmillan, 2014. DOI: 10.1057/9781137406897.0011.

DOI: 10.1057/9781137406897.0011

The second combination, (bottom-left) called 'distributed capitalism', matches distributed control over the back-end while maintaining a focus on capital accumulation. Under this technological regime, P2P infrastructures are designed in such a way as to allow the autonomy and participation of many players. Any Commons is a by-product or after-thought of the system, and personal motivations are driven by exchange, trade and profit. Various P2P developments can be seen within this context, striving for a more inclusionary, distributed and participative capitalism. Though they can be considered part of, say, an anti-systemic entrepreneurialism directed against monopolies and predatory inter-mediaries, they retain the focus on profit making. In the first scenario of netarchical capitalism, control and governance are located within a single proprietary hierarchy, whereas in distributed capitalism, control is located in the network of participating for-profit entrepreneurs and individuals. While netarchical capitalism mainly exploits human coop-eration, distributed capitalism is premised on the idea that everybody can trade and exchange; or, to put it bluntly, that 'everyone can become an independent capitalist'. Of course, as we already discussed, this idea could be central to a few netarchical projects as well, such as Airbnb and TaskRabbit, which enable the monetization of small players. However, their back-ends are not distributed as with distributed capitalist projects, or in other words, in the anarcho-capitalist/libertarian projects.

The libertarian political ideology, on which many projects from this quadrant are premised, advocates the elimination of the state in favor of individual sovereignty, private property and free/open markets (for a treatise of anarcho-capitalism, see Stringham, 2007). As the following analysis of well-known projects from this technological regime will show, the aforementioned ideology is illusionary. In theory you have equipotential individuals (i.e., everyone can potentially participate in a project), but in practice what one gets is concentrated capital and centralized governance. Moreover, we see the emergence of oligarchies and aristocracies. One could postulate that the anarcho-capitalist design of this technological regime, based on the Austrian school of econom-ics (see Schulak and Unterköfler, 2011), in many ways exacerbates the characteristics of the neoliberal era. As stated above, the P2P currency Bitcoin and the Kickstarter crowdfunding platform are representative examples of these developments.

To begin with, Bitcoin was first introduced in 2008 in a paper by Satoshi Nakamoto, which is presumed to be a pseudonym. It is basically

DOI: 10.1057/9781137406897.0011

a FLOSS (i.e., part of the Commons) that supports the movement of currencies. The software circumvents banks and enables the circulation of alternative currency by exploiting P2P networks. Instead of distributing the currency through a centralized network controlled by a central bank, Bitcoins are distributed by nodes participating in a P2P network (much like the BitTorrent file-sharing protocol). Further, as a FLOSS, the Bitcoin system can be monitored by all users worldwide, while participants in the development and improvement of its code cannot make changes that transcend the logic of its original design. Bitcoin is often viewed as an 'apolitical currency' (Varoufakis, 2013), devoid of the troubles that burden other currencies due to it simply being code which is controlled by no one. Yet this is not the case. Besides the fact that there are signs of emerging governance structures in Bitcoin, we can also see that its entire logic follows the key rules of other currencies. The code is in charge rather than the central banks, but as Lessig (2006) puts it, on the Internet the 'code is law', thus pointing out the politicalness imbued in each piece of software. In the real world, the law enables banks to mediate credit transactions between various parties. The law ensures the credibility of contracts, protects property rights and regulates money circulation (Lessig, 2006). Whereas in the digital world, according to Lessig (2006), the code assumes this role and defines what users can and cannot do. Therefore Bitcoin, as a piece of software, is imbued with ideas drawn from a certain political framework, as explained earlier. In other words, the P2P aspect of this project is actually not in the people, but in the computer and the code.

Moreover, Bitcoin is deliberately scarce. By limiting it to 24 million units, Nakamoto (or whoever is behind this project) has created a condition in which the more popular Bitcoin becomes, the higher its price gets. Of course, this makes it more and more difficult to use. The buyer will be motivated to stall any transactions to take advantage of the climbing price, while the seller, for instance an artisan, would buy material now and, by the time the final product is ready, the price would be unfavorable. In short, a deflationary currency puts pressure on the producer/seller to sell as fast as possible, while buyers prefer to wait in order to maximize their purchases. This situation clearly leads to crises. Presumably, the creators' intention was to create a currency free from debt, in the spirit of various politico-economical critiques against the credit system. Bitcoins do not come about as credit relations between two parties but rather as 'private' information in a network.

DOI: 10.1057/9781137406897.0011

The formulation of a Bitcoin 'aristocracy' is the result of the code's architecture. Members of this aristocracy are those that got into the Bitcoin game early on, when it was easy to create new units, as well as the owners of the so-called monster machines, powerful computers that specialize in Bitcoin mining (Davies, 2013). This small percentage of users have accumulated a great deal of Bitcoins, thus not only exhibiting features of the credit system it is supposed to be trying to overcome, but also threatening the viability of the whole project. Our thesis is that Bitcoin is not a Commons-oriented project aiming to satisfy the needs of society, rather a currency that inaugurates distributed capitalism. This new iteration of capitalism conforms to the characteristics of the network era and utilizes P2P infrastructures in order to achieve capital accumulation. Bitcoin is designed to allow multiple users, providing autonomy, but in a competitive framework. It might appear that it exists outside the financial system but, by promoting scarcity and competition, this project aggravates the overaccumulation of capital and exacerbates the social inequalities that it is supposed to combat.

Furthermore, Kickstarter is a crowdfunding platform which enables people to pledge money to provide the means for projects to happen. If the money is raised, the project is then funded, and the people who pledged get whatever they were promised. Kickstarter functions as a reverse market with prepaid investment. In other words, it can be seen as an extension of capitalism: instead of going to the banks for money, you go to the people. According to our four-scenario approach and depending on the point of view, Kickstarter could be considered a netarchical project as well. However, since the surplus value that is extracted here comes from the P2P financing of each project, and thus, the back-end coincides with the front-end (at least from the users' perspective), we place Kickstarter in the bottom-left quadrant, although quite near to the upper-left quadrant (Figure 5.1).

According to Bulajewski (2012), Kickstarter is actually a sophisticated web-hosting provider which charges '60 times the actual cost of providing a service by skimming a percentage off financial transactions'. In other words, Bulajewski (2012) concludes, '[Kickstarter] is the very definition of parasitic capitalism.' He is not the only one who considers Kickstarter as scam, pinpointing its exploitative nature. One could find hundreds of similar allegations and critiques online, but only a few scholarly papers on the topic. Setting these accusations aside, it remains a fact that Kickstarter is nothing more than a web-hosting provider with

DOI: 10.1057/9781137406897.0011

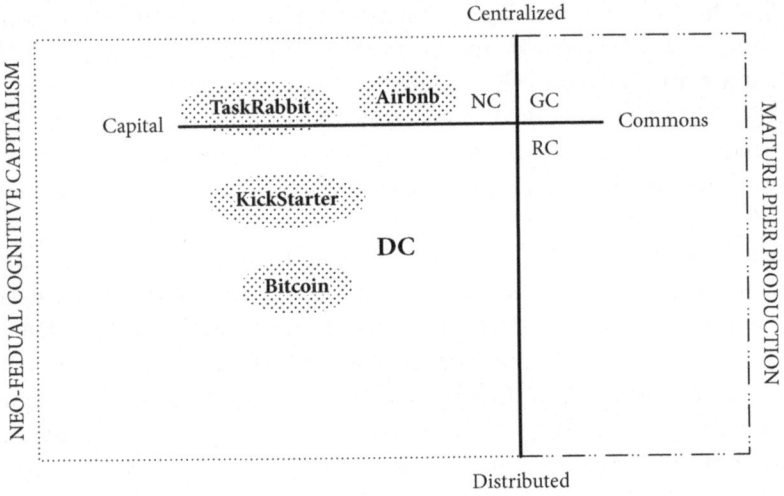

FIGURE 5.1   *The distributed capitalism quadrant*

an exchange platform and no community aspect, although it carries some interesting dynamics. This argument becomes more evident if we look at Kickstarter versus the community-oriented funding platform Goteo, whose projects must have a strong connection to the Commons. In the next part, we will address more projects such as Goteo; however, it is important first to highlight the progressive aspects, if any, of the cognitive capitalist projects that were already brought to the fore. This will be discussed in the following chapter.

DOI: 10.1057/9781137406897.0011

6

The Social Dynamics of the Mixed Model of Neo-feudal Cognitive Capitalism

Abstract: *This chapter highlights the progressive aspects of the cognitive capitalist projects that were discussed in the two previous chapters. The critical observation and documentation of the potentialities of such projects can arguably offer valuable lessons and chances for utilization. This chapter claims that Commons-based communities can benefit from capitalist platforms while struggling for their own rights as the real value creators and could potentially take over such platforms as common or publicly owned utilities. Moreover, they can fork and/or utilize best practices and technologies developed with a for-profit orientation.*

Kostakis, Vasilis and Michel Bauwens. *Network Society and Future Scenarios for a Collaborative Economy.* Basingstoke: Palgrave Macmillan, 2014. DOI: 10.1057/9781137406897.0012.

We argued that the mixed model of neo-feudal cognitive capitalism, as described through the two scenarios/technological regimes of the left quadrants, creates some untenable contradictions, such as a crisis of value. Moreover, we saw that the two scenarios of the emerging value model of cognitive capitalism share two characteristics in principle: first, their main aim is profit-maximization; and second, whatever social goods or relations might be produced are subsumed to the profit model. Hence, it becomes necessary to imagine a transition to a model where the relations of production will not be in contradiction with the evolution of the mode of production and the orientation will rest on the Commons. However, we realize that many forms of the first two scenarios are hybrid because they also allow the further growth of P2P sociality, in which media exchange and production are widely available to an ever-larger user base. For instance, platforms such as Facebook, YouTube or Twitter could become a social utility. The instrumental role of proprietary social media in the success of, for example, the Egyptian anti-government protests which led to the resignation of Egypt's dictatorial leader is almost unquestionable (Eltantawy and Wiest, 2011; Khamis and Vaughn, 2011; Vargas, 2012). Or imagine a YouTube owned by filmmakers and cinemas, and an Amazon owned by authors and independent bookshops. Therefore, there are netarchical platforms that build P2P infrastructures and create some positive conditions which should be critically utilized for a more autonomous network society. Another example is IBM and its coalition with various Commons-based projects in the realm of software. As already postulated, IBM profits on the use value produced through peer production processes. Nevertheless, its involvement has catalyzed the enhancement of the outputs and contributed to the sustainability of many Commons-based projects offering chances for paid labor.

In addition, moving to the distributed capitalism scenario, Bitcoin is extremely important as a signpost, since it is the first global 'post-Westphalian' currency based on 'social sovereignty'. It actually shows that alternative currencies could scale and exist as a workable alternative. Bitcoin, whether it will fail to meet its ambitious goals or not, has paved the way for a new type of currency that utilizes new technological infrastructures, and whose dynamics should not be ignored. As discussed, Bitcoin's protocol enables a decentralized network to achieve consensus, without requiring any trust between parties. Also, the potential of its innovations (e.g., the blockchain) is so big that it has caught the attention of major banking institutions. However, we would say that the most

DOI: 10.1057/9781137406897.0012

important achievement is that it envisions an alternative approach to tackling the major problems in the current credit system. As an open source software program, Bitcoin can get upgraded and it can also get forked. We are witnessing a plethora of new digital currencies based on Bitcoin which aim to surpass some of the issues that were discussed in the previous chapter. Their efforts revolve around the belief that the current financial system is based on an unsustainable principle of continuous growth, and attempt to implement social values into their structure. Openmoney and the OpenUDC are indicative of such efforts. Both projects provide the opportunity for communities to create their own alternative currencies. Peercoin, on the other hand, functions similarly to Bitcoin, but attempts to overcome its problems. Some of these currencies are based on the trust between members of a community of producers and consumers, while others allow mathematics to eliminate the concept of interest from the core of the financial system. Furthermore, crowdfunding platforms, such as Kickstarter, have sometimes enabled the funding and the development of novel, Commons-oriented projects. For example, at the time of this writing (March 2014), more than 220 projects have been tagged as open source and a considerable number of those have been successfully funded, according to the Kickstarter (2014) website.

The critical observation and documentation of the potentialities of projects placed in the two left quadrants can offer valuable lessons and opportunities for utilization. Commons-based communities can benefit from capitalist platforms while struggling for their own rights as the real value creators and, in conditions of social strength, could potentially take over such platforms as common or publicly owned utilities. Moreover, they can fork and/or utilize best practices (e.g., the case of Goteo in relation to Kickstarter) and technologies (e.g., the Bitcoin protocol) developed with a for-profit orientation. We propose that this can happen through the creation of non-capitalist, community-supportive, benefit-driven entities that participate in market exchange without participating in capital accumulation. Before articulating some preliminary policy proposals for such a working hypothesis, we should discuss the next two scenarios/technological regimes based on a different orientation, that of building, empowering and protecting the Commons sphere.

DOI: 10.1057/9781137406897.0012

Part III
The Hypothetical Model of Mature Peer Production: Toward a Commons-Oriented Economy and Society

Plenty of attention has been gathering around the Commons (see Ostrom, 1990; Hardt and Negri, 2011; Barnes, 2006; Benkler, 2006; Bollier and Helfrich, 2012). But what is its concept all about? As we will discuss below, echoing Bollier (2014), the Commons might simultaneously refer to shared resources, a discourse, a new/old property framework, social processes, an ethic, a set of policies or, in other words, to a paradigm of a pragmatic new societal vision beyond the dominant capitalist system. To begin with, in general Commons refers to shared resources where each stakeholder has an equal interest (Ostrom, 1990). The Commons sphere can include natural gifts such as air, water, the oceans and wildlife, and shared 'assets' or creative work such as the Internet, the airwaves, the languages, our cultural heritage and public knowledge which have been accumulating since time immemorial (Bollier, 2002, 2005, 2009). The Commons, with a capital 'C' to highlight its (re)emergence as a powerful counterweight to government and corporate power, also includes goods that have been developed and maintained jointly by a community (Siefkes, 2012; Mackinnon, 2012). These

DOI: 10.1057/9781137406897.0013

goods are shared according to certain community-defined rules (Siefkes, 2012). Take, for example, the Wikipedia encyclopedia or FLOSS, with regard to certain community-driven governance mechanisms through which these projects have managed to remain sustainable, functional and productive. Therefore, it could be said that every Commons scheme basically has four interlinked components: a resource (material and/ or immaterial; replenishable and/or depletable); the community which shares it (the users, administrators, producers and/or providers); the use value created through the social reproduction or preservation of these common goods; and the rules and the participatory property regimes that govern people's access to it. There is an interplay among the afore-mentioned components and, therefore, as we discuss below, Commons should mostly be viewed as social processes.

In contrast to the traditional understanding of property, a key characteristic of the Commons is that no one has exclusive control over the use and disposition of any particular resource (Benkler, 2006). Unlike most things in modern capitalist society, the Commons is neither private nor public, in the traditional sense (The Ecologist, 1994, p. 109). The Commons may signify the absence of state, corporate and/or individual control, in favor of distributed control based upon non-exclusionary, P2P property regimes (Boyle, 2003a, 2003b; Bauwens, 2005). It would be interesting, here, to address the relation between the definitions of the public domain and the Commons. Both concepts are often used interchangeably, yet the latter seems to overtake the former in terms of popularity (Boyle, 2003a, 2003b). The public domain concept is related to the 'outside' of the IP system; it entails items free of property rights, and, thus, emphasizes totally open accessibility: nobody is excluded and everything is allowed (Boyle, 2003a). On the other side, the Commons can be restrictive in a sense. For instance, some Commons-based projects give the freedom to use and/or modify the resource under the condition that new contributions will also be open to others under the same conditions. Hence, the Commons is not an ungoverned space but rather a legal regime for ensuring that the artifacts of community-based productive efforts remain under the control of that community: 'The GPL, the CC licenses, databases of traditional knowledge, and sui generis national statutes for protecting biological diversity all represent innovative legal strategies for protecting the commons' (Bollier, 2009, p. 219). Therefore, we may consider the public domain as a container in which the Commons represents its content of jointly held resources (Ciffolilli,

DOI: 10.1057/9781137406897.0013

2004). When Hardin was discussing the tragedy of the Commons in his 1968 essay, he was actually describing a regime free of property rights and/or of governance mechanisms, where everybody could take and use anything with no constraint. However, in the Commons, a distinct community of users governs the resource (Bollier, 2014, p. 3). Hardin's thesis has also been called 'The Tragedy of Unmanaged, Laissez-Faire, Common-Pool Resources with Easy-Access for Non-Communicating, Self-Interest Individuals' (Hyde, 2010, p. 44). We do not argue that humans are not self-interested and competitive beings, but that they simultaneously exhibit deep concern for fairness, communication, reciprocity, solidarity and social connection: 'all these human traits', Bollier (2014, p. 3) writes, 'lie at the heart of the commons'. Benkler (2011) brings empirical evidence to the fore and describes how cooperation in Commons-based projects triumphs over self-interest, making a case against the blind adherence to 'free market' dogmas.

On the one hand, the neoliberal economics have integrated both the state and the market into one organism/entity, the 'market/state', which stresses the 'deep interdependencies among large corporations, political leaders, and government bodies' (Bollier, 2014, p. 1). The market/state rarely takes into consideration any 'positive' human trait when designing and implementing public policies. Rather, it sees competition, individualism and private property as key drivers of growth and innovation. A critique against neoliberalism could be that it systematizes only a very limited aspect of complex human nature. In contrast, the P2P-driven, Commons-oriented social systems are designed not for one motivation (rational self-interest), but for a multitude of motivations (it is motivation-agnostic). No matter how 'selfish' is the motivation of the Linux or Wikipedia contributors, the system is designed to ensure that participating individuals contribute to the Commons. In the narrow sense, it could be said that the P2P-driven, Commons-based production efforts encapsulate complex human behavior so that it can contribute to the creation of Commons.

Moreover, the mainstream economic theory and many of its prominent indexes (such the Global National Product, GDP) are incapable of recognizing the value produced through various Commons-based projects. Typically, the Commons-oriented forms of production do not produce commodities, but rather use value, and, thus, the latter is not treated as property (The Ecologist, 1994). Hence, the Commons is not recognized as having economic value and cannot take part in

DOI: 10.1057/9781137406897.0013

market exchange within its social/collective/non-exclusive format (Brown, 2010). To tackle this problem, the capitalist political economy would treat the shared resource as a commodity. Enclosed by a certain exclusive property regime – property is a political institution, as Brown (2010) points out – the resource can now enter the market and become a means for profit maximization. According to this perspective, wealth is synonymous with the accumulation of properties; therefore, everything has to be commodified, even things that are more than commodities:

> Labor is only another name for a human activity which goes with life itself, which in its turn is not produced for sale but for entirely different reasons, nor can that activity be detached from the rest of life, be stored or mobilized; land is only another name for nature, which is not produced by man; actual money, finally, is merely a token of purchasing power which, as a rule, is not produced at all, but comes into being through the mechanism of banking or state finance. None of them is produced for sale. The commodity description of labor, land, and money is entirely fictitious. (Polanyi, 1944/2001, pp. 75–76)

From the parliamentary enclosures in England (15th–19th centuries) to the recent 'corporate enclosures', a vast range of commonly held resources has been enclosed, privatized, traded in the market, and thus abused (Bollier, 2002; McCann, 2012). The first wave of enclosure forced people who had been making their living outside the wage mechanism to leave their lands for the cities, where they began to be dependent on wages for their survival (Brown, 2010). They became workers, cogs of the capitalist mode of production. If we follow Marx (1992/1885, 1993/1973), this was an alienation of the self from itself, because what workers produced was very divorced from who they were, thus damaging their essential integrity. And as Brown (2010, p. 120) remarks: 'the alienation of labor caused by an economics of property has repeated itself with a vengeance in our relationship with the living planet.'

However, in the second wave of enclosure, taking place nowadays, there is a robust counter-power: the distributed movement of the Commons with a local and global orientation. There are areas where the market is retreating, not to the bureaucracies and command structures, but instead to the Commons (Stadler, 2014): from seed-sharing cooperatives, the FLOSS and open hardware communities, to localities that use alternative currencies, resilient communities and movements such as community-supported agriculture and Transition Towns. We are observing a re-emergence and flowering of new economic forms based

DOI: 10.1057/9781137406897.0013

on equity, including the cooperative economy, the social economy and the solidarity economy. The reduction of transaction and coordination costs through the modern ICT and the distribution of productive capital in the form of networked personal computers have strengthened this current and given birth to new forms of production based on the collaborative efforts of autonomous individuals. These collaborative modes of social production, which principally celebrate open access to knowledge, have mainly been labeled 'Commons-based peer production' (see Benkler, 2006, 2011; Bauwens, 2005, 2009). The first Commons-based peer production (CBPP) projects were observed in the sphere of information economy, where the marginal cost of information production is very low, if not nearly zero (Benkler, 2006; Bauwens, 2005; Rifkin, 2014; Kostakis, 2012). A plethora of projects, such as the development of the Linux Kernel, the Apache Web server, the office suite LibreOffice, the browser Mozilla Firefox, and the operating system Ubuntu, and free/open content projects such as the encyclopedia Wikipedia, exemplify the productive and governance processes of CBPP. Moreover, we have observed similar patterns of production in some emerging or even not-so-new Commons-based, P2P projects in the primary and secondary economic sector.

As a first example, the Centre for Sustainable Agriculture in India is a community-managed agriculture model that focuses on developing and promoting locally adapted and sustainable farming systems. It was developed to provide a viable alternative for Indian farmers who were being crushed by the cost of chemical pesticides, fertilizers and genetically modified seeds. Open source seed-sharing networks and community seed banks have been set up to overcome the various IP limitations that turned seeds, traditionally considered a Commons, into objects of exclusionary property (Dafermos, 2014). These efforts aim to create a knowledge database (an agricultural Commons, one might say) for the conservation and revival of existing varieties as well as for practices of participatory plant breeding on a local basis (Aoki, 2009; Kloppenburg, 2010; Raidu and Ramanjaneyulu, 2008). Moreover, several producer-consumer cooperatives have been set up with their own meeting grounds (Dafermos, 2014). Another P2P project that goes beyond the information sphere of production is the Transition Towns movement, a grassroots network of communities that is working to build resilience in response to peak oil, climate destruction, and economic instability. Its approach is based on 'a concisely crafted

DOI: 10.1057/9781137406897.0013

methodology for catalyzing community participation via a messy open source organizational process' (Robb, 2009). Likewise, the Open Source Ecology project concerns the development of several low-cost machines meant to cover all sorts of agricultural, and even manufacturing, needs. The design information for these machines is globally available under Commons-oriented licenses adapted for hardware. Another initiative of great interest is the RepRap project, which initially included the development of a low-cost open source 3D printer that could replicate itself by printing a number of its own components. Its lack of IP restrictions has enabled a huge community to experiment with, and improve on, the design. As a result, several models based on the first RepRap model have recently been developed. In addition, not only multiple start-ups but also some large companies began making low-cost 3D printers based on the RepRap design. Another example of CBPP efforts in the manufacturing sector would be the Wikispeed project. Its aim is to produce an energy efficient and modular car made at a fraction of the price of a conventional car. Developed by an international community of volunteers, the Wikispeed car can be built on demand in micro-factories with the use of free/open source software and hardware. Anyone can use or contribute to the project, as all of the specifications are available.

Under the lens of a processual vision of social change (Papadopoulos, Stephenson and Tsianos, 2008), these socially driven projects could be considered as escape routes to alternative forms of social organization. If the political agenda for a world driven by social-oriented values should include the removal of property relations as the economy's foundation and their replacement with civic relation, or, access to resources over ownership (Brown, 2010), then the Commons-oriented movement seems to be emblematic of the aforementioned approach. IP rights are reconfigured to prevent the monetization and expropriation of knowledge. New institutionalized licenses have been introduced to allow the unobstructed sharing of information, including the Creative Commons licenses, the General Public Licenses (GPL) or the Peer Production Licenses (PPL). These forms of property allow the social reproduction of Commons-oriented projects. In other words, knowledge is considered a common good and becomes available to anyone through the utilization of the Internet. Thus, experimentation, collaborative innovation and development are truly promoted while remaining community-driven (Moglen, 2004; Wendel de Joode, 2005; Benkler, 2006).

DOI: 10.1057/9781137406897.0013

However, the aforementioned projects, which form what we could call 'the hypothetical model of mature peer production under civic dominance' (right quadrants), may differ in their focus on the Commons as either local or global. We use the term 'local' as a space distinct from the larger regional, national and international spaces (Sharzer, 2012). In addition, local can be also relational, seen as a moment in the global capital accumulation (ibid.). On the other side, the use of the term 'global' recognizes the possibility that a project might be local, but with the meaning of a spatial territory. This is to say, a project can be rooted somewhere, but the produced use value is principally aimed at a global audience. Our main idea is that networks are global-local, thus, for-benefit orientations can either focus on pure relocalization strategies (though they can be globally organized to achieve this), or they can take a global perspective and create global Commons through global for-benefit associations and global entrepreneurial coalitions. In the 'resilient communities' (RC) scenario (bottom-right) there is distributed control over the P2P infrastructures, that is, both the back-end and the front-end are solely distributed. The focus here is mostly on relocalization and the re-creation of local communities. It is often based on an expectation of a future marked by severe shortages or, in any case, increased scarcity of energy and resources, and so takes the form of lifeboat strategies. Initiatives such as the Transition Towns movement, the degrowth movement or certain aspects of the India-based CSA can be seen in that context. The 'Global Commons' (GC) approach (upper-right) is in contrast to the aforementioned focus on the local, focusing instead on the global Commons. Advocates and builders of this scenario argue that the Commons should be created and fought for on a transnational global scale. The necessity to scale up the Commons is evident in this particular scenario. As becomes obvious, contrary to the left quadrants we do not deal here in terms of technological regimes. Instead, we are more interested in the orientation that communities and individuals have when utilizing P2P infrastructures. The following chapters discuss separately and in more detail each scenario, concluding with some transition proposals for moving toward a global Commons-oriented economy, which arguably can take full advantage of the current TEP's potential in a more sustainable and just way.

DOI: 10.1057/9781137406897.0013

7
Resilient Communities

Abstract: *This chapter addresses the third future scenario which has a local orientation with a focus on the Commons. In the 'resilient communities' scenario there is distributed control over the P2P infrastructures while the focus is mostly on relocalization and the re-creation of local communities. It is often based on an expectation for a future marked by severe shortages of energy and resources, and it often takes the form of lifeboat strategies. However, the resilient communities do not build global structures when the issue, according to Kostakis and Bauwens, is how to organize a global counter-power that can propose alternative modes of social organization on a global scale.*

Kostakis, Vasilis and Michel Bauwens. *Network Society and Future Scenarios for a Collaborative Economy*. Basingstoke: Palgrave Macmillan, 2014. DOI: 10.1057/9781137406897.0014.

The primarily ecological and subsequently economic, social, cultural and political crises the world is facing is the point of departure for the resilient communities approach. This scenario contains strategies and policies for strengthening the ability to adapt to such uneven changes. It makes the case for a transition to a low-carbon, sustainable sharing economy based on social justice and cooperative interactions between people, where economic growth is out of the picture (Lewis and Conaty, 2012). For instance, the degrowth movement along with the Transition Towns, the car sharing and the general permaculture movements can be seen in this context (Figure 7.1).

The theoretical bedrock of the degrowth movement is the so-called degrowth economics, associated with the work of Latouche (2009). According to this body of thought, a radical shift has to take place from growth as the main objective of the modern economy toward its opposite, that is, contraction and downshifting (Foster, 2011; Latouche, 2009). Latouche's work has since given rise to new intellectual movements and inspired a revival of radical Green thought, especially in Europe, as manifested by some prominent conferences in Paris (2008) and Barcelona (2010) (Foster, 2011). The Transition Towns movement, among others, has been influenced by the ideas of degrowth economics. The goal here is the radical relocalization of politics, economics

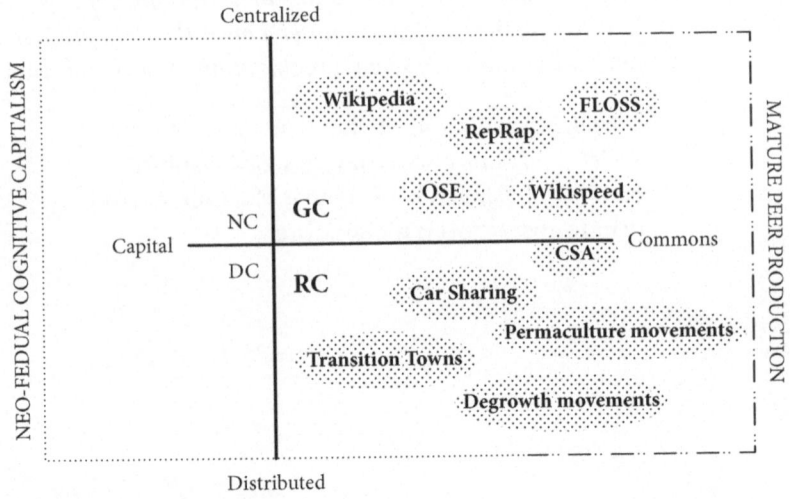

FIGURE 7.1   *The Commons-oriented quadrants*

DOI: 10.1057/9781137406897.0014

and culture to autonomous and self-sufficient communities, in order to become resilient to mega changes, such as peak oil and climate change. Hopkins – who, in 2006, created a working model of a Transition Town community in Totnes, UK – first introduced this concept in his 2008 book *The Transition Handbook*. Since then there have been over a hundred networked transition communities in existence or in the planning stages (see Chamberlin, 2009; Hopkins, 2011). Such communities are of a size that would allow members to have a strong personal influence over collective decisions (Hopkins, 2008, 2011). The Transition Towns concept has as its bedrock not only open source organizational practices, but also the principles of permaculture in combination with resilience and relocalization. Permaculture, a term which stands for 'permanent agriculture', is the design and maintenance of agricultural ecosystems which have the diversity, stability and resilience of natural ecosystems (Mollison, 1988). As Mollison (1988, p. ix–x) puts it:

> The philosophy behind permaculture is one of working with, rather than against nature; of protracted and thoughtful observation rather than protracted and thoughtless action; of looking at systems in all their functions rather than asking only one yield of them; and of allowing systems to demonstrate their own evolutions.

A system based on permaculture principles and practices can evolve, self-organize and thereby survive almost any change: there is no insistence on a single culture, which would shut down learning and cut back resilience (Meadows, 2008, p. 160). Hence, in order to counter the volatility and fragility of the dominant system, building resilience locally is fundamental (Lewis and Conaty, 2012). It is vital to shift to a system with the capacity 'to evolve without losing its core sense of identity or purpose' (Wilding, 2011, p. 19). Therefore, resilience can be seen as the degree to which the system is capable of learning, self-organizing and adapting while remaining coherent (Carpenter et al., 2001; Walker et al., 2009; Folke, 2006). Walker and Salt (2006) along with Lewis and Conaty (2012) highlight some key aspects of any system's resilience: diversity, modularity (consisting of components which can independently operate and be modified), reciprocity, social capital (i.e., trust and bond among members) and tightness of feedback loops. In general:

> [A] system's resilience is enhanced by more diversity and more connections, because there are more channels to fall back on in times of trouble or change. Efficiency, on the other hand, increases through streamlining,

DOI: 10.1057/9781137406897.0014

which usually means reducing diversity and connectivity... Because both are indispensable for long-term sustainability and health, the healthiest flow systems are those that maintain an optimal balance between these two opposing pulls. (Walker and Salt, 2006, p. 121)

Steps and policies toward the world that the resilient communities' scenario envisions can be: the support of a dynamic local economy; the empowerment of local governance and local control; the optimization of assets; the valuing of local distinctiveness and of permaculture; the development of sustainable infrastructures (e.g., affordable housing; interest-free banks; community land trusts; autonomous energy production etc.); and the construction of a social solidarity economy (Wilding, 2011; Lewis and Conaty, 2012).

The local focus of the resilient communities quadrant becomes, however, evident. In extreme forms, this scenario contains simple lifeboat strategies and initiatives, aimed at the survival of small communities in the context of generalized chaos. They may build on the idea that we must accept the reality of considerably more expensive energy and food (Lewis and Conaty, 2012). What marks some of these initiatives is arguably the abandonment of the ambition of scale while the feudalization of territorial integrity is considered mostly inevitable. Though global cooperation and web presence may exist, the focus remains on the local. Most often, political and social mobilization at scale is seen as not realistic, and doomed to failure. In the context of our profit making versus Commons axis though, these projects are squarely aimed at generating community value. We consider them a healthy reaction against global problems and environmental degradation.

Resilient communities try to be immune to the dominant system and they use P2P practices and technologies for good reasons. They try to support individuals' physical and psychological well-being by generating a positive sense of place, localizing the economy within ecological limits and securing entrepreneurial/community stewardship of the local Commons (Wilding, 2011). They do not, however, build global structures. According to our understanding, the issue is how to organize a global counter-power able to propose alternative modes of social organization on a global scale. For Sharzer (2012), 'localism' is the fetishization of scale, as some positive benefit is ascribed to a place precisely because it is small. He argues that resilient communities and other similar projects inevitably become parts of the broader capitalist economy, because they do not confront capitalism, but rather avoid it. Initiatives such as

DOI: 10.1057/9781137406897.0014

Transition Towns are growing movements, though with local focus. They can coexist in harmony within the next scenario of global Commons by the logic that whatever is heavy is local (e.g., desktop manufacturing technologies), and whatever is light is global (e.g., global knowledge Commons).

In addition to the focus on the local, the degrowth narrative is central to the resilient communities scenario. We believe, quoting Foster (2011), 'that the ecological struggle, understood in these terms, must aim not merely for degrowth in the abstract but more concretely for de-accumulation – a transition away from a system geared to the accumulation of capital without end'. To realize such a transition it is crucial to develop pragmatic alternatives. Similar to how we began talking about 'alter-globalization' when the 'antiglobalization' movement became counter-intuitive, we now need to become more positive and start talking about 'alter-growth' scenarios instead of thinking in anti-growth/degrowth terms. Arguably, the issue is not to produce and consume less per se, but to develop new models of production which will work on a higher level than capitalist models. We consider it difficult to challenge the dominant system if we lack a working plan to transcend it. A post-capitalist world is bound to entail more than a mere reversal to pre-industrial times. As the TEPS theory informs us, the adaptation of current institutions and the creation of new ones take place in the deployment phase of each TEP. We claim that the times are, finally, mature enough to introduce a radical political agenda with brand new institutions, fueled by the spirit of the Commons and aiming to provide a viable global alternative to the capitalist paradigm beyond degrowth or antiglobalization rhetorics.

DOI: 10.1057/9781137406897.0014

8
Global Commons

Abstract: *This chapter deals with the 'Global Commons' scenario which celebrates the hypothetical model of mature peer production. Advocates of this scenario argue that the Commons should be created and fought for on a global scale. Though production is distributed and therefore facilitated at the local level, the resulting micro-factories are considered as essentially networked on a global scale, profiting from the mutualized global cooperation both on the design of the product, and on the improvement of the common machinery. Political and social mobilization, on regional, national and transnational scale, is seen as part of the struggle for the transformation of institutions. According to Kostakis and Bauwens, this scenario does not take social regression as given, and believes in sustainable abundance for the whole of humanity.*

Kostakis, Vasilis and Michel Bauwens. *Network Society and Future Scenarios for a Collaborative Economy*. Basingstoke: Palgrave Macmillan, 2014. DOI: 10.1057/9781137406897.0015.

DOI: 10.1057/9781137406897.0015

Several global-oriented Commons-based projects such as FLOSS, Wikipedia, Wikispeed, RepRap or Open Source Ecology (OSE) highlight the emergence of technological capabilities shaped by human factors, which in turn shape the environment in which humans live and work. They create what Benkler (2006, p. 31) calls new 'technological-economic feasibility spaces' for social practice. These feasibility spaces include different social and economic arrangements, where profit, power and control do not seem as predominant as they have in the history of modern capitalism. From this new communicational, interconnected, virtual environment, a new social productive model is emerging, different from the industrial one. We are witnessing the emergence of a new proto-mode of production, that is, Commons-based peer production, based on distributed, collaborative forms of organization. It is developing within capitalism, rather as Marx (1979) argued that the early forms of merchant and factory capitalism developed within the feudal order. In other words, system change is back on the agenda, but in an unexpected form, not as a 'socialist' alternative, but as a Commons-based alternative. As we saw, capitalism in its present form is facing limits, especially resource limits, and in spite of the rapid growth of the BRICS (Brazil, Russia, India, China and South Africa) economies, it is undergoing a process of decomposition. The question is whether the new proto-mode can generate the institutional capacity and alliances needed to break the political power of the old order. Ultimately, the potential of the new mode is the same as those of previous proto-modes of production – to emancipate itself from dependency on the old decaying mode, to become self-sustaining and thus replace the accumulation of capital with the circulation of the Commons. In an independent circulation of the Commons, the common use value would directly contribute to the further strengthening of the Commons and of the commoners' own sustainability, without dependence on capital. How could this be achieved? Before dealing with this tempting question, we believe that it is crucial to shed more light on the social, economic and political dynamics of CBPP.

When it comes to information, CBPP is more productive than market-based or the 'bureaucratic-state' systems (Benkler, 2006). It produces social well-being because it is based on people's intrinsic positive motivations (i.e., the need to create, learn and communicate) and synergetic cooperation among participants and users (Benkler 2006; Hertel, Niedner, and Herrmann, 2003; Lakhani and Wolf, 2005). As

DOI: 10.1057/9781137406897.0015

Hertel, Niedner and Herrmann (2003, p. 1174) mention in their study of the incentives of 141 Linux kernel community participants, the latter were driven 'by similar motives as voluntary action within social movements such as the civil rights movement, the labor movement, or the peace movement'. Benkler (2006) makes two intriguing economic observations which challenge some 'eternal truths' of the mainstream economic theory. Commons-based projects fundamentally challenge the assumption that in economic production, the human being solely seeks profit maximization. Volunteers contribute to information production projects, while they gain knowledge, experience and reputation, and communicate with each other motivated by intrinsically positive incentives. This does not mean that the monetary motive is totally absent; however, it is relegated to a peripheral concept (Benkler, 2006). The second challenge is directed against the conventional wisdom that, in Benkler's (2006, p. 463) words, 'we have only two basic free transactional forms – property-based markets and hierarchically organized firms.' CBPP can be considered a third way, and should not be treated as an exception but rather as a widespread phenomenon, although it is not currently counted in the economic census (Benkler, 2006). In terms of neoliberal economics, what is happening in CBPP can arguably be considered only in the sense that individuals are free to contribute, or take what they need, following their individual inclinations, with an invisible hand bringing it all together yet without any monetary mechanism. Hence, in contrast to markets, in CBPP the allocation of resources is not done through a market-pricing mechanism. Hybrid modes of governance are employed, and what is generated is not profit, but a Commons.

CBPP is based on practices that stand in contrast to those of the market-based business firms. More specifically, CBPP is opposed to industrial firms' hierarchical control and authority. Instead, it is based on communal validation and negotiated coordination (see, for instance, Dafermos' (2012) study on the Free BSD project's collectivist and consensus-oriented governance system) as quality control is community-driven, and conflicts are solved through an ongoing mediated dialogue (e.g., in Wikipedia, the dialogue takes place in the discussion page of each article). However, in cases such as the internal battle between inclusionists and deletionists, Wikipedia's lack of a clearly defined constitution led a small number of participants to create rules in conflict with others: persistent, well-organized minorities adroitly handled their opponents, seriously challenging the sustainability of the project (Kostakis, 2010). Therefore, it must be stressed that when abundance is replaced by scarcity (as happened in Wikipedia when deletionists demanded strict content control), power structures emerge because CBPP mechanisms cannot function well (Kostakis, 2010). Investigating prominent CBPP projects, O'Neil (2009) analyzed the tensions generated by the distribution of authority, and showed that it is important to discuss openly how power and authority actually work in CBPP in order to be able to organize differently. His proposal is that leaders must support maximum autonomy for participants toward a more egalitarian situation. Of course, a special characteristic of CBPP is that if these benevolent dictators (Kostakis, 2010) abuse their power, their leadership becomes malicious, and a substantial exodus of community members often occurs. These members, due to the low marginal costs of information, are free to start their own new project, using the already Commons-based peer-produced information if they wish.

Further, CBPP is not driven by the for-profit orientation that defines market projects, as peer projects have a for-benefit orientation, creating use value for their communities. This does not mean that the profit motive is totally absent in CBPP projects, but rather that incentives such as learning, communication and experience come to the fore. That is how the human person actually operates, rather than the imagined homo economicus. Besides, Hess' (2005, p. 515) 'private-sector symbiosis' hypothesis outlines that emphasis on technology and product innovation can lead 'to the articulation of social movements goals with those of inventors, entrepreneurs, and industrial reformers' (2005, p. 516). Therefore, 'a cooperative relationship emerges between advocacy organizations that support the alternative technologies/products, and private sector firms that develop and market alternative technologies' (ibid.). For instance, Linux and IBM come in accordance with Hess' argument for the private-sector symbiosis and subsequent incorporation and transformation of the technologies which may, though, provoke, an object conflict. 'As the technological/product field undergoes diversification', Hess (2005, p. 515) writes, object conflicts 'erupt over a range of design possibilities, from those advocated by the more social movement-oriented organizations to those advocated by the established industries'. It can be claimed that an object conflict is taking place concerning the Makerbot Replicator 2 3D printer, which is partly closed source. This may, arguably, lead to the loss of Makerbot's community (Giseburt, 2012).

DOI: 10.1057/9781137406897.0015

Instead of the division of labor in CBPP, a distribution of modular tasks takes place, with anyone able to contribute to any module, while the threshold for participation is as low as possible (see Benkler, 2006; Bauwens, 2005; Tapscott and Williams, 2006; Dafermos and Söderberg, 2009). Modularity is a key condition for CBPP to emerge: 'Described in technical terms, modularity is a form of task decomposition. It is used to separate the work of different groups of developers, creating, in effect, related yet separate sub-projects' (Dafermos and Söderberg, 2009, p. 61). Torvalds (1999), the instigator of the Linux project, maintains that the Linux kernel development model requires modularity, because in that way, people can work in parallel. Empirical research (see MacCormack, Rusnak and Baldwinet, 2007; Dafermos, 2012) shows that modular design is characteristic not just of Linux but of the FLOSS development model in general. According to Carson (2010, p. 208) 'The Unix philosophy of providing lots of small specialized tools that can be combined in versatile ways is probably the oldest expression in software of this modular style.' We also observe the same approach in the development of one of the most prominent CBPP projects, namely Wikipedia. Articles (i.e., modules), which consist of sections (or, sub-modules), are built upon other articles and entries produced, and thus can be used individually as well as in combination. By breaking up the raw elements into smaller modules, there is both an abundance of options in terms of remixing them, as well as a low participation threshold, since the individuals can have access to the modules rather than centralized forms of capital. Further, modularity leads to stigmergic collaboration. In its most generic formulation, according to Marsh and Onof (2007, p. 1), 'stigmergy is the phenomenon of indirect communication mediated by modifications of the environment.' Therefore, in the context of CBPP, stigmergic collaboration is the 'collective, distributed action in which social negotiation is stigmergically mediated by Internet-based technologies' (Elliott, 2006).

Moreover, CBPP is opposed to the rivalry (scarcity of goods) through which market profit is generated, as sharing the created goods does not diminish the value of the good, but actually enhances it (Benkler, 2006). To this, one might add that CBPP is facilitated by free, unconstrained and creative cooperation of communities, which lowers the legal restrictive barriers to such a process and invents new, institutionalized ways of sharing. In terms of property, as we have discussed, the Commons is an idea different both from state property, where the state manages a certain resource on behalf of the people, and from private property, where

DOI: 10.1057/9781137406897.0015

a private entity excludes the common use of it. It is, however, important to highlight that the contributors of CBPP projects do have interests and rights concerning their work and are interested in protecting their intellectual property (O'Mahony, 2003). Thus, the Commons-oriented approach to property 'does not assert that sharing is an ethical absolute' (after all everyone is, or should be, free to choose what type of license they will adopt), but tries to balance the rights of innovators with the rights of the public (O'Mahony, 2003; von Hippel and von Krogh, 2003). It becomes obvious that what sets CBPP apart from the proprietary-based mode of production – the 'industrial one' (Benkler, 2006) – is its modes of governance (consensus-oriented governance mechanisms) and property (communal shareholding), whose foundation stones are the abundance of resources, openness and the power of meaningful human cooperation. These are the very characteristics of CBPP which provide the capacity to deliver genuinely innovative, remarkable results (thus contesting allegations of low quality: see Keen, 2007; Lanier, 2010) such as the Apache web server, Mozilla Firefox browser, Linux kernel, BIND (the most widely used DNS software), Sendmail (router of the majority of e-mail) and a myriad of emerging open source hardware projects.

Of course, beyond the great potential of CBPP, there may well be numerous obstacles, theoretical and practical problems, and negative side effects. However, taken in this idealized context, CBPP arguably carries some aspects which create a political economy where economic efficiency, profit and competitiveness cease to be the sole guiding stars (Moore and Karatzogianni, 2009), while civil society attains a more important role, bringing (back) the notion of the Commons into the heart of the economy (Orsi, 2009). Under these lenses, the Commons can be seen as a legitimate vehicle of citizenship or as an equivalent of Tocqueville's (2010) civil society, through which citizens mobilize and express their interests while protecting their rights (Mackinnon, 2012). It can be central to the process of civilizing the economy, which would require a strong notion of citizenship – of membership in a global civil society (Brown, 2010). The Commons movement is removing property relations as our political economy's foundation and is replacing them with civic relations that define our bonds with each other – at work, in neighborhoods, in cities and in global communities (ibid.). The Commons is long-term social and material processes that cannot be created overnight: 'in order to become meaningful they must exist over an extensive period of time' (Stadler, 2014, p. 31). In other words, the

DOI: 10.1057/9781137406897.0015

various spheres of the Commons are products of P2P creative processes as they expand horizontally and in dense interconnections with each other. Therefore, we must go beyond a material understanding of the concept and approach the Commons not only as a resource or as a property regime, but mainly as a social process. Producing a categorization or taxonomy of the Commons by a type of resource can be misleading, as Bollier (2014) warns us:

> While choosing to categorize commons by the type of resource involved is tempting, a focus on the resource alone can be misleading. For example, a 'knowledge commons' on the Internet is not simply about intangible resources such as software code or digital files; such a commons also requires physical resources to function (computers, electricity, food for human beings). By the same token, 'natural resource commons' are not just about timber or fish or corn, because these resources, like all commons, can only be managed through social relationships and shared knowledge.

In other words, to quote Helfrich (2013), 'all commons are social, and all commons are knowledge commons'. Our relationships to shared goods that are managed as Commons should be the focal point and, thus, we should discuss the process of Commoning. In other words, we should discuss the process of the circulation of the free/open/participatory: 'free' and 'open' ensure access to raw material to build the Commons; 'participatory' refers to the process of broad participation in order to actually build it. The Commons, then, becomes the institutional format used to prevent private appropriation of shared creations, and the circle is closed when Commons-generated material is once again free/open raw material for the next circulation of the Commons.

The 'Global Commons' approach (upper-right) focuses on a larger scale in relation to the resilient communities quadrant, that is, on the Commons with a global orientation (Figure 7.1). Advocates and builders of this scenario argue that the Commons should be created and fought for on a transnational global scale. Though production is distributed and therefore facilitated at the local level, the conjunction of CBPP with desktop manufacturing technologies could create sustainable business ecologies. There, the resulting micro-factories, essentially networked on a global scale, would profit from mutualized global cooperation, both on the design of the product and on the improvement of common machinery. 'Micro-factories' is a concept that refers to small dimension, automated factories capable of greatly conserving resources such as space, energy, materials and time (Tanaka, 2001; Okazaki, Mishima

DOI: 10.1057/9781137406897.0015

and Ashida, 2004). They are likely to feature automatic machine tools, assembly systems, evaluation and control systems, a quality inspection system and waste elimination system (Kussul et al., 2002; Koch, 2010). For example, see the Wikispeed's project micro-factory in Seattle, which is a licensed light-industrial space the size of a shipping container, used as a prototyping facility for cars that can get more than 100 miles per gallon (Denning, 2012). The Wikispeed car is produced voluntarily by a network of developers from all over the world, who have managed to significantly reduce the development time and cost compared with conventional car manufacturing, through the use of methods similar to those of CBPP (Dafermos, 2014; Denning, 2012). The Wikispeed project was launched in the 2008 Progressive Insurance Automotive X-PRIZE competition for the development of energy-efficient cars (Dafermos, 2014). The resolution to apply CBPP development methods to car manufacturing was what separated this project from its competition (ibid.). When the founder of this project, Joe Justice, posted his plans on the Web, volunteers gathered and shortly after, a functioning prototype was presented (Denning, 2012; Halverson, 2011). More than 150 volunteers contribute now, and their goal is to deliver Wikispeed as a complete car for $17,995 and as a kit for $10,000 (Wikispeed, 2012). To sum up, as Dafermos (2014) puts it, Wikispeed, just like that of Open Source Ecology and RepRap projects, demonstrates how a technology project can leverage the open design Commons and P2P infrastructures to engage the global community in its development. Most importantly, Wikispeed suggests a model of distributed manufacturing that is well-suited to a post-fossil fuel economy: a model which is small scale ('on-demand'), decentralized, energy efficient and locally controlled (Dafermos, 2014).

Any distributed enterprise, such as the ones being developed around the aforementioned projects, is seen in the context of transnational 'phyles', that is, alliances of ethical enterprises that operate in solidarity around a particular knowledge Commons (P2P Foundation, 2014; de Ugarte, 2014). As the key terrain of conflict is around the relative autonomy of the Commons vis-à-vis for-profit companies, we are in favor of a preferential choice toward entrepreneurial formats which integrate the value system of the Commons, rather than profit maximization. In that context, phyles, in other words the creation of businesses by the community, can make the Commons viable and sustainable in the long run. Advocates and builders of this scenario struggle for a shift from the current flock of community-oriented businesses toward business-enhanced

DOI: 10.1057/9781137406897.0015

communities. They believe that we need corporate entities which are sustainable from the inside out, not just via external regulation from the state, but from their own internal statutes and links to Commons-oriented value systems. We are arguably living the endgame of neoliberal material globalization based on cheap energy, which necessitates relocalization of production (see the resilient communities scenario). However, we have new possibilities for online, affinity-based socialization, coupled with the resulting physical interactions and community building. The value-creation communities of this quadrant might be locally based but are globally linked. Out of that, new forms of business organization may arise, which are substantially more community-oriented. This scenario sees no contradiction between global open design collaboration and local production: both can occur simultaneously, so the relocalized reterritorialization will be accompanied by global tribes, organized in phyles. The various Commons, based on shared knowledge, code and design, will be part of these new global knowledge networks, but closely linked to relocalized implementations.

Therefore, political and social mobilization on the regional, national and transnational scale is seen as part of the struggle for the transformation of institutions. Participating enterprises are vehicles for the commoners to sustain Global Commons as well as their own livelihoods. This scenario does not take social regression as a given and believes in sustainable abundance for the whole of humanity. It envisions a transition to a paradigm which would include new decentralized and distributed systems of provisioning and democratic governance, escaping the pathologies of the current political economy and constructing an ecologically sustainable alternative (Bollier, 2014). To achieve such a transition, the Global Commons scenario suggests that we should work on building both global and local political and social infrastructures. Next, we venture into some general transition proposals for the state and the market in order to realize the full potential of the ICT-driven TEP in a more sustainable and just way.

DOI: 10.1057/9781137406897.0015

9

Transition Proposals toward a Commons-Oriented Economy and Society

Abstract: *There is arguably a need for political and social mobilization on regional, national and transnational scale, with a political agenda that would transform people's expectations, the economy, the infrastructures and the institutions in the vein of a Commons-oriented political economy. According to Kostakis and Bauwens, the latter is not a utopia or just a project for the future. Rather it is rooted in an already existing social and economic practice. This chapter concludes with some transition proposals for moving toward a global Commons-oriented economy which can take full advantage of the current techno-economic paradigm's potential in a more sustainable and just way.*

Kostakis, Vasilis and Michel Bauwens. *Network Society and Future Scenarios for a Collaborative Economy*. Basingstoke: Palgrave Macmillan, 2014. DOI: 10.1057/9781137406897.0016.

In the midst of the current techno-economic transformations, humanity is at a crossroads. How will a degraded natural environment sustain a political economy based on the assumption that natural resources are an endless sink? How will the modern, participatory ICT be fine-tuned, with the assumption that potentially abundant cultural/knowledge resources would exist in artificial scarcity? What value models will be adopted for a deployment period to come? Which model will prevail? According to Brynjolfsson and McAfee (2011) 'When the changes happen faster than expectations and/or institutions can adjust, the transition can be cataclysmic'. To avoid such a cataclysm, we arguably need political and social mobilization on the regional, national and transnational scale, with a political agenda that would transform our expectations, our economy, our infrastructures and our institutions in the vein of a Commons-oriented political economy. The latter is not a utopia or simply a project for the future. Rather, it is rooted in an already existing social and economic practice, that of the CBPP, which is producing Commons of knowledge, code and design, and has created real economies such as the FLOSS economy, the open hardware economy and others. In its broadest interpretation, concerning all the economic activities emerging around open and shared knowledge, it has increasingly been contributing trillions of dollars to the GDP of the USA, according to the Fair Use Economy report (Rogers and Szamosszegi, 2011) (and one should reckon how difficult it is for the GDP index to consider socially produced use value).

We have already described the micro-economic structures of this emerging Commons-oriented economic model, which we can summarize as follows: at the core of this new value model are contributory communities, consisting of both paid and unpaid labor, which are creating common pools of knowledge, code and design. These contributions are enabled by collaborative infrastructures of production, and a supportive legal and institutional infrastructure, which enables and empowers the collaborative practices. These infrastructures of cooperation, that is technical, organizational and legal infrastructures, are very often enabled by democratically run foundations. These foundations are more generically called 'for-benefit associations', which may create code/design/knowledge depositories; protect against infringements of open and sharing licenses; organize fundraising drives for infrastructure; and organize knowledge sharing through local, national and international conferences. Thus, they are an enabling and protective mechanism. Finally, successful projects

DOI: 10.1057/9781137406897.0016

create an economy around the Commons pools, based on the creation of added value products and services that are based on the common pools, but also add to them. This is done by entrepreneurs and businesses that operate in the marketplace. Most often, these are for-profit enterprises, creating an 'entrepreneurial coalition' around the Commons and the community of contributors. They hire developers and designers as workers, create livelihoods for them and support the technical and organizational infrastructure, also including the funding of foundations. On the basis of this generic micro-economic experiences, it is possible to deduce adapted macro-economic structures as well, which would include a civil society that consists mainly of communities of contributors creating shareable Commons; of a new state form, which would enable and empower social production generally and create and protect the necessary civic infrastructures; and an entrepreneurial coalition which would conduct commerce and create livelihoods (Figure 9.1).

If we look at the micro-level, we recommend the intermediation of cooperative accumulation. In today's FLOSS economy we have a paradox: the more 'communist' the sharing license we use (i.e., no restrictions

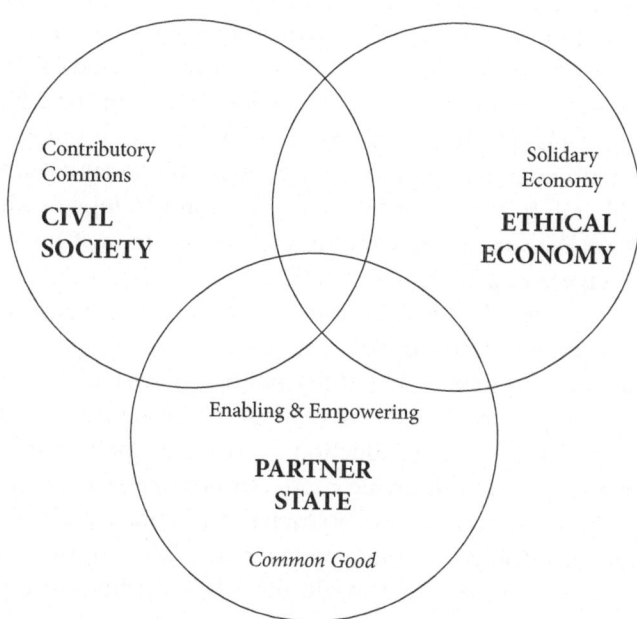

FIGURE 9.1   *The Commons-oriented economic model of mature peer production*

DOI: 10.1057/9781137406897.0016

on sharing) in the peer production of free software or open hardware, the more capitalist the practice (i.e., multinationals can use it for free). Take, for example, the Linux Commons which has become a corporate Commons as well, enriching big, for-profit corporations. It is obvious that this works in a certain way and seems acceptable to most free software developers. But is this way optimal? Indeed, the GPL and its variants allow anyone to use and modify the software code (or design), as long as the changes are integrated back in the commons pool under the same conditions for further users. Our argument does not focus on the legal, contractual basis of the GPL and similar licenses, but on the social logic that they enable, which is: it allows anybody to contribute, and it allows anybody to use. In fact, this relational dynamic is technically a form of 'communism': from each according to his/her abilities, to each according to his/her needs. This paradoxically allows multinational corporations to use free software code for profit maximization and capital accumulation. The result is that we do have an accumulation and circulation of information Commons, based on open input, participatory processes and Commons-oriented output; but it is subsumed to capital accumulation. Therefore, it is not currently possible, or at least easy, to have social reproduction (i.e., to create sustainable livelihoods) within the sphere of the Commons. The majority of the contributors participate on a voluntary basis, and those who have an income make a living either through wage labor or alliances with capital-driven entities. Hence the free software and culture movements, however important they might be as new social forces and expression of new social demands, are also, in essence, 'liberal' in the tradition of the political ideology of liberalism. We could say they are liberal-communist and communist-liberal movements, which create a 'communism of capital'.

The question is whether Commons-based peer production, that is, a new proto-mode of production, can generate the institutional capacity and alliances needed to break the political power of the old order. Ultimately, the potential of the new mode is the same as those of the previous proto-modes of production – to emancipate itself from its dependency on the old decaying mode, to become self-sustaining and thus replace the accumulation of capital with the circulation of the Commons. This would be an independent circulation of the Commons, where the common use value would directly contribute to the further strengthening of the Commons and of the commoners' own sustainability, without dependence on capital. How could this be achieved? Is there

DOI: 10.1057/9781137406897.0016

an alternative? We believe that there is: to replace the non-reciprocal licenses, that is those which do not demand a direct reciprocity from its users, with one based on reciprocity. We argue that the Peer Production License (PPL), designed and proposed by Kleiner (2010), exemplifies this line of argument. PPL should not to be confused with the Creative Commons (CC) non-commercial (NC) license, as its logic is different. The CC-NC offers protection to individuals reluctant to share, as they do not wish a commercialization of their work that would not reward them for their labor. Thus the CC-NC license stops further economic development based on this open and shared knowledge, and keeps it entirely in the not-for-profit sphere. The logic of the PPL is to allow commercialization, but on the basis of a demand for reciprocity. It is designed to enable and empower a counter-hegemonic reciprocal economy that combines a Commons that is open to all that contribute, while charging a license fee for the for-profit companies who would like to use it without contributing. Not that much changes in practice for the multinationals; they can still use the code if they contribute, as IBM does with Linux. However, those who do not contribute should pay a license fee – a practice they are used to. Its practical effect would be to somehow direct a stream of income from capital to the Commons, but its main effect would be ideological, or if you like, value-driven.

The entrepreneurial coalitions that are linked around a PPL-based Commons would be explicitly oriented toward their contributions to the Commons, and the alternative value system that it represents. From the point of view of the peer producers or commoners, a Commons-based reciprocal license, such as PPL, would allow the contributory communities to create their own cooperative entities. In this new ecology, profit would be subsumed to the social goal of sustaining the Commons and the commoners. Even the participating for-profit companies would consciously contribute under a new logic. This proposal would link the Commons to an entrepreneurial coalition of ethical market entities (co-ops and other models) and keep the surplus value entirely within the sphere of commoners/cooperators, instead of leaking out to the multinationals. In other words, through this convergence (or rather combination) of a Commons model for abundant immaterial resources, and a reciprocity-based model for the 'scarce' material resources, the issue of livelihoods and social reproduction could be solved. The surplus value would be kept inside the Commons sphere itself. The cooperatives, through their cooperative accumulation, would fund the production of

DOI: 10.1057/9781137406897.0016

immaterial Commons, because they would pay and reward the peer producers associated with them. In this way, peer production could move from a proto-mode of production, unable to perpetuate itself on its own outside capitalism, to an autonomous and real mode of production. It would create a counter-economy that could be the basis for reconstituting a 'counter-hegemony' with a for-benefit circulation of value. This process, allied to 'pro-Commons' social movements, could be the basis for the political and social transformation of the political economy. Hence we might move from a situation in which the communism of capital is dominant, to a situation in which we have a 'capital for the Commons', increasingly insuring the self-reproduction of the peer-production mode.

The new open cooperativism would be substantially different from the previous form. In the old one, internal economic democracy is accompanied by participation in market dynamics on behalf of the members, using capitalist competition. There is an unwillingness to share profits and benefits with outsiders, therefore, no creation of the Commons. We argue that an independent Commons-oriented economy would need a different model in which the cooperatives produce Commons and are statutorily oriented toward the creation of the common good. To realize their goals they should adopt multistakeholder forms of governance which would include workers, users-consumers, investors and the concerned communities. Today we have a situation where open communities of peer producers are largely oriented toward the start-up model and are subsumed to profit maximization, while the cooperatives remain closed, use exclusive intellectual property licenses, and, thus, do not create a Commons. In the new model of open cooperativism, a merger should occur between the open peer production of the Commons and the cooperative production of value. The new open cooperativism would (i) integrate externalities; (ii) practice economic democracy; (iii) produce Commons for the common good; (iv) and socialize its knowledge. The circulation of the Commons would be combined with the process of cooperative accumulation, on behalf of the Commons and its contributors. In the beginning, the immaterial Commons field, following the logic of free contributions and universal use for everyone who needs it, would co-exist with a cooperative model for physical production, based on reciprocity. But as the cooperative model would become more and more hyper-productive through its ability to create sustainable abundance in material goods, the two logics could merge.

DOI: 10.1057/9781137406897.0016

It is important to highlight that the Commons-based reciprocal licenses, such as PPL, are not merely about redistribution of value, but about changing the mode of production. Our approach is to transform really existing peer production, which today is not a full mode of production, being incapable of assuring its own self-reproduction. This is exactly why the convergence of peer production in the sphere of abundance must be linked to the sphere of cooperative production, to ensure its self-reproduction. As with past phase transitions, the existence of a proto-counter-economy and the resources that this allocates to the counter-hegemonic forces are absolutely essential for political and social change. This was arguably the weakness of classic socialism, in that it had no alternative mode of production and could only institute state control after a takeover of power. In other words, it is difficult, if not impossible, to wait and see the organic and emergent development of peer production into a fully alternative system. If we follow such an approach, peer production would just remain a parasitic modality dependent on self-reproduction through capital. We argue that the expectation that one can change society merely by producing open code and design, while remaining subservient to capital, is a dangerous pipe dream. Through the ethical economy surrounding the Commons, by contrast, it becomes possible to create non-commodified production and exchange. We thus envision a resource-based economy which would utilize stigmergic mutual coordination through the gradual application of open book accounting and open supply chain. We believe that there will be no qualitative phase transition merely through emergence, but that it will require the reconstitution of powerful political and social movements which aim to become a democratic polis. And that democratic polis could indeed, through democratic decisions, accelerate the transition. It could take measures that obligate private economic forces to include externalities, thereby ending infinite capital accumulation.

However, such changes at the level of the micro-economy might not survive a hostile capitalist market and state without necessary changes at the macro-economic level (Kostakis and Stavroulakis, 2013). We should not ignore the fact that the state has its own interests in perpetuating its bureaucracy and legitimacy. Gajewska (2014) emphasizes this argument through the case of the campus food services (free lunches) at Concordia University as an example of peer production in the physical world. She describes the tension between the university administration and the P2P food services collectives which were producing food Commons. The

DOI: 10.1057/9781137406897.0016

project started with 'direct action' occupying university space for cooking, eventually recognized by Concordia University. What we realize is that a transition narrative should take into account the possibility for creating spaces of democratic accountability from below. For example, in the aforementioned case, the university was the framework through which students could pool resources in the form of fee levies and organize for-benefit projects (Gajewska, 2014). Hence, there is a need for transition proposals carried by a resurgent social movement that embraces new value creation through the Commons, and becomes the popular and political expression of the emerging social class of peer producers and commoners. This movement should arguably be allied with the forces representing both waged and cooperative labor, independent Commons-friendly entrepreneurs and agricultural and service workers.

To begin with, we introduce the concept of the Partner State Approach (PSA), in which the state becomes a 'partner state' and enables autonomous social production. The PSA could be considered a cluster of policies and ideas whose fundamental mission is to empower direct social-value creation, and to focus on the protection of the Commons sphere as well as on the promotion of sustainable models of entrepreneurship and participatory politics. It is important to emphasize that we consider the 'partner state' as the ideal condition for a government to pursue (as is the case in Ecuador with the FLOK society project) and the P2P movement to fight for. While people continue to enrich and expand the Commons, building an alternative political economy within the capitalist one, by adopting a PSA the state becomes an arbiter, retreating from the binary state/privatization dilemma to the triarchical choice of an optimal mix among government regulation, private-market freedom and autonomous civil-society projects. Thus, the role of the state evolves from the post-World War II welfare-state model, which could arguably be considered a historical compromise between social movements for human emancipation and capitalist interests, to the partner state one, which embraces win-win sustainable models for both civil society and market. In such an approach, the state would strive to maximize openness and transparency while it would systematize participation, deliberation and real-time consultation with the citizens. Thus, the social logic would move from ownership-centric to citizen-centric. The state should de-bureaucratize through the commonification of public services and public-Commons partnerships. Public service jobs could be considered a common pool resource, and participation could be extended to the

DOI: 10.1057/9781137406897.0016

whole population. Furthermore, representative democracy would be extended through participatory mechanisms (participatory legislation, participatory budgeting etc.). It would also be extended through online and offline deliberation mechanisms as well as through liquid voting (real-time democratic consultations and procedures, coupled with proxy voting mechanisms). In addition to this, taxation of productive labor, entrepreneurship and ethical investing, as well as taxation of the production of social and environmental goods should be minimized. However, taxation of speculative unproductive investments, taxation on unproductive rental income and taxation of negative social and environmental externalities should be augmented. In these ways, the partner state would sustain civic Commons-oriented infrastructures and ethical Commons-oriented market players, reforming the traditional corporate sector in order to minimize social and environmental externalities. Last but not least, of great importance would be the engagement of the partner state in debt-free public monetary creation while supporting a structure of specialized complementary currencies.

The second component of a Commons-oriented economy would be an ethical market economy, that is, the creation of a Commons-oriented social/ethical/civic/solidarity economy. Ethical market players would coalesce around the Commons of productive knowledge, eventually using peer production and Commons-oriented licenses to support the social-economic sector. They should integrate common good concerns and user-driven as well as worker-driven multistakeholders in their governance models. Ethical market players would move from extractive to generative forms of ownership, while open, Commons-oriented ethical company formats are privileged. They should create a territorial and sectoral network of 'chamber of Commons' associations to define their common needs and goals and interface with civil society, commoners and the partner state. With the help from the partner state, ethical market players would create support structures for open commercialization, which would maintain and sustain the Commons. Ethical market players should interconnect with global productive Commons communities (i.e., open design communities) and with global productive associations (phyles) which project ethical market power on a global scale. We suggest that ethical market players should adopt a 1–8 wage differential and minimum and maximum wage levels. The mainstream commercial sector should be reformed to minimize negative social and environmental externalities, while incentives which aim for a convergence between the corporate and solidarity economy must be provided. Hybrid economic forms, such as fair trade and social entrepreneurship, could be encouraged to obtain such convergence. Distributed micro-factories for (g)localized manufacturing on demand should be created and supported in order to satisfy local needs for basic goods and machinery. Institutes for the support of productive knowledge should also be created on a territorial and sectoral basis. Education should be aligned with the co-creation of productive knowledge in support of the social economy and the open Commons of productive knowledge. Therefore, all publicly funded research and innovation should be released under the GPL (for an extensive discussion of this proposal, see Boldrin and Levine's (2013) as well as Pearce (2012)). Additionally, Commons infrastructures for both immaterial and material goods have to be created: in such a political economy, society is seen as a series of interlocking Commons supported by an ethical market economy and a partner state that protects the common good and creates supportive civic infrastructures. Local and sectoral Commons would create civil alliances of the Commons to interface with the chamber of the Commons and the partner state. Interlocking for-benefit associations (knowledge Commons foundations) would enable and protect the various Commons. In addition to this, solidarity cooperatives should form public-Commons partnerships in alliance with the partner state, while the ethical economy sector could be represented by the chamber of Commons. Also, the natural Commons should be managed by a public-Commons partnership and based on civic membership in Commons trusts.

We would like to stress that this list of transitional strategies and preliminary proposals for policymaking is general and non-inclusive. By no means does this chapter intend to formulate a specific economic plan or a clearly defined transitional policy to a Commons-based society. It is important to remember Bouckaert and Mikeladze's (2008, p. 7) advice that 'a more sophisticated diagnosis, as a function of culture, context, and systems features' allows for 'selective transfers, for inspiration by other good practices, for adjustments of solutions, for facilitated learning by doing, for trajectories which are fit for purpose'. Hence, a fundamental belief on which this book is premised is the fact that there are no universal 'how-to' manuals, because not only does every nation have its own special characteristics, but also rapid social change based on grandiose systemic substitutions usually has disastrous results, as history shows; many times these results are contradictory

DOI: 10.1057/9781137406897.0016

to what ambitious but benevolent revolutionaries may struggle for. Therefore, this chapter is an attempt to introduce suggestions and ideas for a post-capitalist society and draw attention to the promising, creative rhetoric of a PSA for Commons-oriented development. We might argue that four factors in a certain state could catalyze the transition toward a Commons-based society: (i) the extended micro-ownership of fixed capital such as land, machinery and so on; (ii) the need for recomposing the productive infrastructures, as is the case in defaulted states; (iii) an already existent robust network of solidarity and cooperative initiatives; (iv) a decentralized energy network. Further interdisciplinary research around these newly developed concepts and ideas on a global basis is imperative, along with initiating a debate between scholars and activists in order to fine-tune the transition scenarios toward Commons-oriented economies and societies.

DOI: 10.1057/9781137406897.0016

Conclusions

Kostakis, Vasilis and Michel Bauwens. *Network Society and Future Scenarios for a Collaborative Economy*. Basingstoke: Palgrave Macmillan, 2014. DOI: 10.1057/9781137406897.0017.

▶

DOI: 10.1057/9781137406897.0017

We discussed three models of value creation, redistribution and economic development:

▸ The classical proprietary capitalism, currently in decline.
▸ The mixed model of cognitive capitalism which is manifested by two different technological regimes/future scenarios. Netarchical and distributed capitalism are aimed at capital accumulation either for the benefit of global shareholders (NC), or for networks of for-profit enterprises and individuals (DC). However, in NC the design of the infrastructure (the back-end) is in the hands of centralized privately owned platforms, whereas in DC the infrastructure is primarily distributed with the promise to make everyone a small capitalist.
▸ The hypothetical mature peer production model whose seeds can be found not only in the global Commons (GC) scenario, but also in the resilient communities (RC) one. They are aimed at improving the circulation of the Commons for the local community (RC) and the transnational Commons (GC). In both scenarios the control is distributed through free self-allocation by the commoners. In the RC the commoners affect the governance and design of their infrastructures on a local scale, whereas in the GC approach the commoners try to build global infrastructures.

Under the conditions of traditional proprietary capitalism we have seen that workers create value in their private capacity as providers of labor. In addition, managerial and engineering layers are introduced in order to manage collective production on behalf of the capitalist owners. The codified knowledge is proprietary and the value is captured as IP rent. The owners of capital capture and realize the market value, whereas there is partial redistribution for the workers in the form of wages. Under conditions of capital-labor balance, the state redistributes wealth to the workers as consumers. However, under the contemporary conditions of labor weakness, the state redistributes wealth to the financial sector and creates conditions of debt dependence for the majority of the population. This value model is becoming obsolete because it contradicts the essential characteristics of the ICT-driven TEP, but is also based on a profoundly counterproductive, twofold logic of social organization. On the one side, this logic stems from a false concept of abundance in the limited material world, since it has created a system based on infinite growth within the confines of finite resources. On the other, it promotes

DOI: 10.1057/9781137406897.0017

a false concept of scarcity in the infinite immaterial world and instead of allowing continuous experimental social innovation, it purposely erects legal and technical barriers to prevent free cooperation through strict copyright, patents and so on. Therefore, the first priority for a sustainable civilization should be transforming these principles into their opposites. We argued that the rise of peer production signals new alternative paths for the deployment of the current TEP. This proto-mode of production is both immanent and transcendent vis-à-vis capitalism, because it has features that strongly decommodify both labor and immaterial value and institute a field of action based on P2P dynamics and a P2P value system. Peer production functions not only within the cycle of accumulation of capital but also within the new cycle of creation and accumulation of the Commons.

The key idea of this book is to distinguish the condition of the P2P/Commons/sharing practices under the dominance of financialized cognitive capitalism, and a more genuine civil/ethical model centered on the Commons. Under conditions of emerging peer production while financial capitalism is still dominant, we saw that civic voluntary contributors, paid labor and independent entrepreneurs create value codified in common pools of knowledge, code and design. The capital owners realize and capture the market value of both contributors and labor, while the proprietary network platforms capture and realize the attention value of the sharers/contributors. The capital owners also profit from the benefits of disaggregated distributed labor (i.e., crowdsourcing). The Commons are managed by for-benefit institutions which reflect the balance of influence between contributors, labor and capital owners, but continue to expand the common pools. However, the Commons sector lacks solidarity mechanisms to cope with precarity and, thus, civil society is still derivate to the market and state sectors. The state weakens its public service and solidarity functions, in favor of its repressive functions as well as subsidizing financial capital. It barely contributes to the co-creation of the conditions for peer production whereas redistribution to financial capital continues.

Under conditions of strong, mature peer production through civic dominance, that is 'genuine' CBPP, we saw that civic voluntary contributors and autonomous cooperative labor would create codified value through common pools. Labor and civic re-skilling could occur through Commons-oriented distributed manufacturing, which places value creators at the helm of distributed manufacturing and other forms of value

DOI: 10.1057/9781137406897.0017

creation. Commons contributors should create cooperative Commons-oriented market entities that sustain the Commons and their communities of contributors. Hence, cooperative and other Commons-friendly market entities would not only co-create common pools but also engage in cooperative accumulation on behalf of their members. Therefore, Commons-oriented contributions should be codified in their legal and governance structures while entrepreneurial coalitions and phyles are formed, meaning structured networks of firms working around joint common pools to sustain Commons-producing communities. Furthermore, the Commons-enabling for-benefit institutions would become a core civic form for the governance of common pools, while the associated market entities would create solidarity mechanisms and income for the peer producers and commoners, supported by the partner state. The state, dominated by the civic/Commons sectors, becomes a partner state which creates and sustains the civic infrastructure necessary to enable and empower autonomous social production. The market becomes a moral and ethical economy, oriented around Commons production and mutual coordination supported by the partner state functions. The market sector is dominated by cooperative, Commons-oriented legal, governance and ownership models, while the remaining profit-maximizing entities are reformed to respect environmental and social externalities.

The hypothetical model of mature peer production can arguably be considered a working alternative which can perform better than the current dominant value model while solving a number of systemic problems. We have attempted to highlight the existence of creative communities who are building the political economy they desire within the confines of the political economy they mean to transcend. Peer production, then, should be seen as a social advancement within capitalism but with various post-capitalistic aspects in need of protection, enforcement, stimulation and connection with progressive social movements. In the midst of a turning point, it is high time we supported a sustainable alternative capable of breaking the shackles of capitalist opportunism and ushering in a new political economy based on the finer aspects of the human spirit. It is high time the accumulation of capital was replaced by a full circulation of the Commons.

DOI: 10.1057/9781137406897.0017

References

Anderson, C. (2012) *Makers: The New Industrial Revolution* (London: Random House).

Aoki, K. (2009) 'Free Seeds, Not Free Beer: Participatory Plant Breeding, OpenSource Seeds, and Acknowledging User Innovation in Agriculture', *Fordham Law Review*, 77(5), 2275–2310.

Arvidsson, A., and Pietersen, N. (2013) *The Ethical Economy: Rebuilding Value after the Crisis* (New York, NY: Columbia University Press).

Baran, P. A., and Sweezy, P. M. (1966) *Monopoly Capital: An Essay on the American Economic and Social Order* (New York, NY: Monthly Review Press).

Barnes, P. (2006) *Capitalism 3.0: A Guide to Reclaiming the Commons* (San Francisco, CA: Berrett-Koehler Publishers).

Bauwens, M. (2005) 'The Political Economy of Peer Production', *CTheory Journal*, http://www.ctheory.net/articles.aspx?id=499, date accessed 11 April 2014.

Bauwens, M. (2009) 'Class and Capital in Peer Production', *Capital & Class*, 33, 121–141.

Bell, D. (1973) *The Coming of Post-Industrial Society* (New York, NY: Basic Books).

Benkler, Y. (2006) *The Wealth of Networks: How Social Production Transforms Markets and Freedom* (New Haven, CT: Yale University Press).

Benkler, Y. (2011) *The Penguin and the Leviathan: The Triumph of Cooperation over Self-Interest* (New York, NY: Crown Business).

DOI: 10.1057/9781137406897.0018

Bessen, J., and Meuer, M. (2009) *How Judges, Bureaucrats, and Lawyers Put Innovators at Risk* (Princeton, NJ: Princeton University Press).

Boldrin, M., and Levine, D. (2007) *Against Intellectual Monopoly* (New York, NY: Cambridge University Press).

Boldrin, Michele, and David K. Levine (2013) 'The Case against Patents', *Journal of Economic Perspectives*, 27(1): 3–22.

Bollier, D. (2002) 'Reclaiming the Commons: Why We Need to Protect Our Public Resources from Private Encroachment', *Boston Review*, 27, 3–4.

Bollier, D. (2005) *Brand Name Bullies: The Quest to Own and Control Culture* (Hoboken, NJ: Wiley).

Bollier, D. (2009) *Viral Spiral: How the Commoners Built a Digital Republic of Their Own* (New York, NY: New Press).

Bollier, D. (2014) 'The Commons as a Template for Transformation', *Great Transition Initiative*, http://www.greattransition.org/document/the-commons-as-a-template-for-transformation, date accessed 11 April 2014.

Bollier, D., and Helfrich, S. (2012) *The Wealth of Commons* (Amherst, MA: Levellers Press).

Bouckaert, G., and Mikeladze, M. (2008) 'Introduction', *The NISPAee Journal of Public Administration and Policy*, 1(2), 7–8.

Boutang, Y. M. (2012) *Cognitive capitalism* (Cambridge: Polity Press).

Boyle, J. (2003a) 'Foreword: The Opposite of Property?' *Law and Contemporary Problems*, 66, 1–32.

Boyle, J. (2003b) 'The Second Enclosure Movement and the Construction of the Public Domain', *Law and Contemporary Problems*, 66, 33–74.

Brown, M. T. (2010) *Civilizing the Economy: A New Economics of Provision* (Cambridge: Cambridge University Press).

Brynjolfsson, E., and McAfee, A. (2011) *Race against the Machine: How the Digital Revolution Is Accelerating Innovation, Driving Productivity, and Irreversibly Transforming Employment and the Economy* (Lexington, MA: Digital Frontier Press).

Bulajewski, M. (2012) 'An Ambitious Plan For Putting Kickstarter Out of Business', *A Blog of Philosophical Reflections & Speculations*, http://www.mrteacup.org/post/an-ambitious-plan-for-putting-kickstarter-out-of-business.html, date accessed 11 April 2014.

DOI: 10.1057/9781137406897.0018

Carpenter, S. R., Walker, B. H., Anderies, J. M., and Abel, N. (2001) 'From Metaphor to Measurement: Resilience of What to What?' *Ecosystems*, 4, 765–781.

Carson, K. (2010) *The Homebrew Industrial Revolution: A Low-Overhead Manifesto* (Charleston, SC: BookSurge Publishing).

Castells, M. (2000) *The Rise of the Network Society* (Oxford: Blackwell).

Castells, M. (2003) *The Power of Identity* (Oxford: Blackwell).

Castells, M. (2009) *Communication Power* (Oxford: Oxford University Press).

Chamberlin, S. (2009) *The Transition Timeline: For a Local, Resilient Future* (Cambridge: Green Books).

Chomsky, N. (2011) *Profit over People: Neoliberalism and Global Order* (New York, NY: Seven Stories Press).

Ciffolilli, A. (2004) 'The Economics of Open Source Hijacking and the Declining Quality of Digital Information Resources: A Case for Copyleft', *First Monday*, 9, http://www.firstmonday.org/ojs/index.php/fm/article/view/1173/1093, date accessed 11 April 2014.

Coleman, B., and Hill, M. (2004) 'How Free Became Open and Everything Else under the Sun', *M/C Journal: A Journal of Media and Culture*, 7, http://journal.media-culture.org.au/0406/02_Coleman-Hill.php, date accessed 11 April 2014.

Dafermos, G. (2012) *Governance Structures of Free/Open Source Software Development* (Delft: Next Generation Infrastructures Foundation).

Dafermos, G. (2014) Policy Paper on Distributed Manufacturing, *FLOK Society Project, Draft policy document*, http://en.wiki.floksociety.org/w/Commons-oriented_Productive_Capacities, date accessed 11 April 2014.

Dafermos, G., and Söderberg, J. (2009) 'The Hacker Movement as a Continuation of Labour Struggle', *Capital & Class*, 33, 53–73.

Davies, K. (2013) The Monster Machines Mining Bitcoins in Cyberspace That Could Make Techies a Small Fortune (But Cost $160,000 a Day to Power), http://www.dailymail.co.uk/news/article-2309673/Techies-building-powerful-computers-Bitcoins-new-digital-currency-make-millions.html, date accessed 11 April 2014.

de Ugarte, D. (2014) *Trilogía de las Redes*, http://lasindias.com/de-las-naciones-a-las-redes, date accessed 11 April 2014.

Denning, S. (2012) 'How Agile Can Transform Manufacturing: The Case of Wikispeed', *Strategy & Leadership*, 40(6), 22–28.

DOI: 10.1057/9781137406897.0018

Drechsler, W., Backhaus, J., Burlamaqui, L., Chang, H.-J., Kalvet, T., Kattel, R., Kregel, J., and Reinert, E. (2006) 'Creative Destruction Management in Central and Eastern Europe: Meeting the Challenges of the Techno-Economic Paradigm Shift' in T. Kalvet & R. Kattel (eds.) *Creative Destruction Management: Meeting the Challenges of the Techno-Economic Paradigm Shift* (Tallinn: Praxis Center for Policy Studies).

Drucker, P. (1969) *The Age of Discontinuity* (London: Heinemann).

Elliott, M. (2006) 'Stigmergic Collaboration: The Evolution of Group Work', *M/C Journal: A Journal of Media and Culture*, 9(2).

Eltantawy, N., and Wiest, J. B. (2011) 'The Arab Spring| Social Media in the Egyptian Revolution: Reconsidering Resource Mobilization Theory', *International Journal of Communication*, 5, http://ijoc.org/index.php/ijoc/article/view/1242/597, date accessed 11 April 2014.

Federici, S., and Caffentzis, G. (2007) 'Notes on the Edu-Factory and Cognitive Capitalism', *The Commoner*, 12, 63–70.

Folke, C. (2006) 'Resilience: The Emergence of a Perspective for Social–Ecological Systems Analyses', *Global Environmental Change*, 16, 253–267.

Foster, J. B. (2011) 'Capitalism and Degrowth: An Impossibility Theorem', *Monthly Review*, 62(8), https://monthlyreview.org/2011/01/01/capitalism-and-degrowth-an-impossibility-theorem, date accessed 11 April 2014.

Freeman, C. (1974) *The Economics of Industrial Innovation* (Harmondsworth: Penguin Books).

Freeman, C. (1996) *The Long Wave in the World Economy* (Aldershot: Edward Elgar).

Fuchs, C., Schafranek, M., Hakken, D., and Breen, M. (2010) 'Capitalist Crisis, Communication, and Culture – Introduction to the Special Issue of TripleC', *TripleC*, 8(2), 193–204.

Fukuyama, F. (1992) *The End of History and the Last Man* (New York, NY: Free Press).

Funnell, W., Jupe, R. E., and Andrew, J. (2009) *In Government We Trust: Market Failure and the Delusions of Privatisation* (London: Pluto Press).

Gajewska, K. (2014) 'Peer Production and Prosummerism as a Model for the Future Organization of General Interest Services Provision in Developed Countries: Examples of Food Services Collectives', *World Future Review*, http://wfr.sagepub.com/content/early/2014/03/07/1946756714522983.abstract, date accessed 11 April 2014.

DOI: 10.1057/9781137406897.0018

Galbraith, J. K. (1993) *A Short History of Financial Euphoria* (New York, NY: Whittle Books).

Giseburt, R. (2012) 'Is One of Our Open Source Heroes Going Closed Source?' *Make*, http://blog.makezine.com/2012/09/19/is-one-of-our-open-source-heroes-going-closed-source/, date accessed 11 April 2014.

Godet, M. (2000) 'The Art of Scenarios and Strategic Planning: Tools and Pitfalls', *Technological Forecasting and Social Change*, 65(1), 3–22.

Gore, A. (2013) *The Future: Six Drivers of Global Change Hardcover* (New York, NY: Random House).

Halverson, M. (2011) 'Wikispeed's 100 Mile Per Gallon Car', *Seattle Met*, http://www.seattlemet.com/issues/archives/articles/wikispeeds-100-mpg-car-january-2011/1, date accessed 11 April 2014.

Hardin, G. (1968) 'The Tragedy of the Commons', *Science, 162*, 1243–1248.

Hardt, M., and Negri, A. (2011) *Commonwealth* (Cambridge, MA: The Belknap Press).

Harvey, D. (2007) *The Limits to Capital* (London: Verso).

Harvey, D. (2010) *The Enigma of Capital: And the Crises of Capitalism* (New York, NY: Oxford University Press).

Harvey, D. (2012) *Rebel Cities: From the Right to the City to the Urban Revolution* (London: Verso).

Helfrich, S. (2013) 'Economics and the Commons? Towards a Commons-Creating Peer Economy', Economics and the Commons Conference, Berlin, http://commonsandeconomics.org/2013/06/09/silke-helfrichs-opening-keynote-towards-a-commons-creating-peer-economy, date accessed 11 April 2014.

Hertel, G., Niedner, S., and Herrmann, S. (2003) 'Motivation of Software Developers in Open Source Projects: An Internet-Based Survey of Contributors to the Linux Kernel', *Research Policy, 32*, 1159–1177.

Hess, D. (2005) 'Technology- and Product-Oriented Movements: Approximating Social Movement Studies and Science and Technology Studies', *Science, Technology, & Human Values, 4*, 515–535.

Hopkins, R. (2008) *The Transition Handbook: From Oil Dependency to Local Resilience* (Cambridge: Green Books).

Hopkins, R. (2011) *The Transition Companion: Making Your Community More Resilient in Uncertain Times* (Cambridge: Green Books).

Howe, J. (2008) *Crowdsourcing: Why the Power of the Crowd Is Driving the Future of Business* (New York, NY: Crown Business).

DOI: 10.1057/9781137406897.0018

Hyde, L. (2010) *Common as Air: Revolution, Art, and Ownership* (New York, NY: Farrar, Straus and Giroux).

IBM (International Business Machines Corporation) (2010) 'IBM Is Committed to Linux and Open Source', *IBM*, http://www-03.ibm.com/linux/, date accessed 11 April 2014.

Kalvet, T., and Kattel, R. (2006) *Creative Destruction Management: Meeting the Challenges of the Techno-Economic Paradigm Shift* (Tallinn: Praxis Center for Policy Studies).

Keen, A. (2007) *The Cult of the Amateur* (New York, NY: Doubleday).

Kelly, R., Sirr, L., and Ratcliffe, J. (2004) 'Futures Thinking to Achieve Sustainable Development at Local Level', *Foresight*, 6(2), 80–90.

Khamis, S., and Vaughn, K. (2011) Cyberactivism in the Egyptian Revolution: How Civic Engagement and Citizen Journalism Tilted the Balance, *Arab Media and Society*, 14, http://www.arabmediasociety.com/?article=769, date accessed 11 April 2014.

Kickstarter (2014) Most Funded Open Source Projects, https://www.kickstarter.com/discover/advanced?tag_id=20&sort=most_funded, date accessed 11 April 2014.

Kleiner, D. (2010) *The Telekommunist Manifesto* (Amsterdam: Institute of Network Cultures).

Kloppenburg, J. (2010) 'Impeding Dispossession, Enabling Repossession: Biological Open Source and the Recovery of Seed Sovereignty', *Journal of Agrarian Change*, 10(3), 367–388.

Koch, M. D. (2010) 'Utilizing Emergent Web-Based Software Tools as an Effective Method for Increasing Collaboration and Knowledge Sharing in Collocated Student Design Teams', *Oregon State University*, (MSc Thesis), http://ir.library.oregonstate.edu/xmlui/handle/1957/16855, date accessed 11 April 2014.

Kondratieff, N. D. (1979) 'The Long Waves in Economic Life', *Review*, 2, 519–562.

Kostakis, V. (2010) 'Peer Governance and Wikipedia: Identifying and Understanding the Problems of Wikipedia's Governance', *First Monday*, 15, http://firstmonday.org/ojs/index.php/fm/article/view/2613, date accessed 11 April 2014.

Kostakis, V. (2012) 'The Political Economy of Information Production in the Social Web: Chances for Reflection on Our Institutional Design', *Contemporary Social Science*, 7, 305–319.

DOI: 10.1057/9781137406897.0018

Kostakis, V., and Stavroulakis, S. (2013) 'The Parody of the Commons', *TripleC*, 11(2), 412–424.

Kostakis, V., Fountouklis, M., and Drechsler, W. (2013) 'Peer Production and Desktop Manufacturing: The Case of the Helix_T Wind Turbine Project', *Science, Technology & Human Values*, 38(6), 773–800.

Kussul, E., Baidyk, T., Ruiz-Huerta, L., Caballero-Ruiz, A., Velasco, G., and Kasatkina, L. (2002) 'Development of Micromachine Tool Prototypes for Microfactories', *Journal of Micromechanics and Microengineering*, 12(6), 795–812.

Lakhani, K., and Wolf, R. (2005) 'Why Hackers Do What They Do: Understanding Motivation and Effort in Free/Open Source Software Projects' in J. Feller, B. Fitzgerald, S. Hissam & K. Lakhani (eds.) *Perspectives on Free and Open Source Software* (pp. 3–22) (Cambridge, MA: MIT Press).

Lanier, J. (2010) *You Are Not a Gadget: A Manifesto* (New York, NY: Knopf).

Latouche, S. (2009) *Farewell to Growth* (Cambridge, MA: Polity).

Leigh, A. (2003) 'Thinking Ahead: Strategic Foresight and Government', *Australian Journal of Public Administration*, 62(2), 3–10.

Lessig, L. (2004) *Free Culture: How Big Media Uses Technology and the Law to Lock Down Culture and Control Creativity* (New York, NY: Penguin Press).

Lessig, L. (2006) *Code Version 2.0* (New York, NY: Basic Books).

Lewis, M., and Conaty, P. (2012) *The Resilience Imperative: Cooperative Transitions to a Steady-State Economy* (Gabriola Island: New Society Publishers).

MacCormack, A., Rusnak, J., and Baldwin, C. Y. (2007) 'The Impact of Component Modularity on Design Evolution: Evidence from the Software Industry', *Harvard Business School Technology & Operations Mgt. Unit, 08-038*, http://ssrn.com/abstract=1071720, date accessed 11 April 2014.

MacKinnon, R. (2012) *Consent of the Networked* (New York, NY: Basic Books).

Marsh, L., and Onof, C. (2007) 'Stigmergic Epistemology, Stigmergic Cognition', *Cognitive Systems Research*, 9(1–2), 136–149.

Marx, K. (1979) *A Contribution to the Critique of Political Economy* (New York, NY: Intl Pub).

Marx, K. (1992/1885) *Capital: Critique of Political Economy* (London: Penguin Classics).

DOI: 10.1057/9781137406897.0018

Marx, K. (1993/1973) *Grundrisse: Foundations of the Critique of Political Economy* (London: Penguin).

McCann, A. (2012) 'Opportunities of Resistance: Irish Traditional Music and the Irish Music Rights Organisation 1995–2000', *Popular Music and Society, 35*(5), 651–681.

Meadows, D. (2008) *Thinking in Systems: A Primer* (Vermont, VT: Chelsea Green Publishing).

Miles, I. (2004) 'Scenario Planning' in UNIDO (ed.) *Foresight Methodologies: Training Module 2.* (Vol. 159, pp. 67–91) (Vienna: UNIDO).

Moglen, E. (2004) 'Freeing the Mind: Free Software and the Death of Proprietary Culture', *Maine Law Review, 56*(1), 1–12.

Mollison, B. (1988) *Permaculture: A Designers' Manual* (Tyalgum: Tagari Publications).

Moore, P., and Karatzogianni, A. (2009) 'Parallel Visions of Peer Production', *Capital & Class, 33*, 7–11.

Morozov, E. (2012) *The Net Delusion: The Dark Side of Internet Freedom* (New York, NY: Public Affairs).

Mueller, M. L. (2010) *Networks and States: The Global Politics of Internet Governance* (Cambridge, MA: MIT Press).

Mulgan, G. (2013) *The Locust and the Bee: Predators and Creators in Capitalism's Future* (Princeton, NJ: Princeton University Press).

Nakamoto, S. (2008) Bitcoin: A Peer-to-Peer Electronic Cash System, http://bitcoin.org/bitcoin.pdf, date accessed 11 April 2014.

Neeson, J. M. (1993) *Commoners: Common Right, Enclosure and Social Change in England, 1700–1820* (Cambridge: Cambridge University Press).

O'Neil, M. (2009) *Cyberchiefs: Autonomy and Authority in Online Tribes* (London: Pluto Press).

O'Mahony, S. (2003) 'Guarding the Commons: How Community Managed Software Projects Protect Their Work', *Research Policy, 32*(1179–1198).

Okazaki, Y., Mishima, N., and Ashida, K. (2004) 'Microfactory – Concept, History, and Developments', *Journal of Manufacturing Science and Engineering, 126*(4), 837–844.

Orsi, C. (2009) 'Knowledge-Based Society, Peer Production and the Common Good', *Capital & Class, 33*, 31–51.

Ostrom, E. (1990) *Governing the Commons: The Evolution of Institutions for Collective Action* (Cambridge: Cambridge University Press).

DOI: 10.1057/9781137406897.0018

P2P Foundation (2014) Phyles, http://p2pfoundation.net/Phyles, date accessed 11 April 2014.

Papadopoulos, D., Stephenson, N., and Tsianos, V. (2008) *Escape Routes: Control and Subversion in the 21st Century* (London: Pluto Press).

Pariser, E. (2011) *The Filter Bubble* (New York, NY: Penguin Viking).

Patry, W. (2009) *Moral Panics and the Copyright War* (New York, NY: Oxford University Press).

Pearce, J. M. (2012) 'Physics: Make Nanotechnology Research Open-Source', *Nature, 491,* 519–521.

Perez, C. (1983) 'Structural Change and Assimilation of New Technologies in the Economic and Social Systems', *Futures, 15,* 357–375.

Perez, C. (1985) 'Long Waves and Changes in Socio-Economic Organizations', *IDS Bulletin, 16*(1), 36–39.

Perez, C. (1988) 'New Technologies and Development' in C. Freeman & B.-A. Lundvall (eds.) *Small Countries Facing the Technological Revolution* (pp. 85–97)(London: Pinter).

Perez, C. (2002) *Technological Revolutions and Financial Capital: The Dynamics of Bubbles and Golden Ages* (Cheltenham: Edward Elgar Pub).

Perez, C. (2009a) 'The Double Bubble at the Turn of the Century: Technological Roots and Structural Implications', *Cambridge Journal of Economics, 34,* 779–805.

Perez, C. (2009b) 'Technological Revolutions and Techno-Economic Paradigms', *Cambridge Journal of Economics, 33,* 185–202.

Polanyi, K. (1944/2001) *The Great Transformation: The Political and Economic Origins of Our Time* (Boston, MA: Beacon Press).

Raidu, D. V., and Ramanjaneyulu, G. (2008) 'Community Managed Sustainable Agriculture' in B. Venkateswarlu, S. S. Balloli & Y. S. Ramakrishna (eds.) *Organic Farming in Rainfed Agriculture: Opportunities and Constraints* (Hyderabad: Central Research Institute for Dryland Agriculture.).

Rifkin, J. (2011) *The Third Industrial Revolution: How Lateral Power Is Transforming Energy, the Economy, and the World* (New York, NY: Palgrave Macmillan).

Rifkin, J. (2014) *The Zero Marginal Cost Society: The Internet of Things, the Collaborative Commons, and the Eclipse of Capitalism* (New York, NY: Palgrave Macmillan).

Robb, J. (2009) 'Transition Towns and Participatory Problem Solving', *Global Guerrillas,* http://globalguerrillas.typepad.com/

DOI: 10.1057/9781137406897.0018

globalguerrillas/2009/04/rc-journal-transition-towns-as-a-means-to-participative-problem-solving.html, date accessed 11 April 2014.

Rogers, T., and Szamosszegi, A. (2011) 'Fair Use in the U.S. Economy: Economic Contribution of Industries Relying on Fair Use', *OER Knowledge Cloud*, https://oerknowledgecloud.org/?q=content/fair-use-us-economy-economic-contribution-industries-relying-fair-use-0, date accessed 11 April 2014.

Schmidt, E., and Cohen, J. (2013) *The New Digital Age : Reshaping the Future of People, Nations and Business* (New York, NY: Alfred A. Knopf).

Schmoller, G. (1898/1893) 'Die Volkswirtschaft, die Volkswirtschaftslehre und Ihre Methode' in G. Schmoller (ed.) *Über einige grundfragen der socialpolitik und der volkswirtschaftslehre* (Berlin: Duncker and Humblot).

Scholz, T. (2012) *Digital Labor: The Internet as Playground and Factory* (New York, NY: Routledge).

Schulak, E. M., and Unterköfler, H. (2011) *The Austrian School of Economics: A History of Its Ideas, Ambassadors, & Institutions* (Auburn, AL: Ludwig von Mises Institute).

Schumpeter, J. (1975/1942) *Capitalism, Socialism and Democracy* (London: Harper and Row).

Schumpeter, J. (1982/1939) *Business Cycles* (Philadelphia, PA: Porcupine Press).

Schwartz, P. (1996) *The Art of the Long View: Planning for Future in an Uncertain World* (New York, NY: Currency Doubleday).

Sharzer, G. (2012) *No Local: Why Small-Scale Alternatives Won't Change the World* (Winchester: John Hunt Publishing).

Siefkes, C. (2012) 'The Boom of Commons-Based Peer Production' in D. Bollier & S. Helfrich (eds.) *The Wealth of Commons* (Amherst, MA: Levellers Press).

Stadler, F. (2014) *Digital Solidarity* (Lüneburg: Mute & PML Books).

Stiglitz, J. (2010) *Freefall: America, Free Markets, and the Sinking of the World Economy* (New York, NY: W.W. Norton).

Stringham, E. (2007) *Anarchy and the Law: The Political Economy of Choice* (Oakland, CA: Independent Institute).

Tanaka, M. (2001) 'Development of Desktop Machining Microfactory', *Riken Review, 34*, http://pdf.aminer.org/000/353/685/development_of_a_micro_transfer_arm_for_a_microfactory.pdf, date accessed 11 April 2014.

DOI: 10.1057/9781137406897.0018

Tapscott, D., and Williams, A. (2006) *Wikinomics: How Mass Collaboration Changes Everything* (New York, NY: Portfolio).

The Ecologist (1994) 'Whose Common Future: Reclaiming the Commons', *Environment and Urbanization,* 6(1), 106–130.

Tocqueville, A. de. (2010) *Democracy in America* (New York, NY: Signet Classics).

Torvalds, L. (1999) 'The Linux Edge' in C. DiBona, S. Ockman & M. Stone (eds.) *Open Sources: Voices from the Open Source Revolution* (pp. 101–109) (Sebastopol, CA: O'Reilly).

van der Heijden, K. (2005) *Scenarios: The Art of Strategic Conversation* (New York, NY: John Wiley & Sons).

van der Heijden, K., Bradfield, R., Burt, G., Cairns, G., and Wright, G. (2002) *The Sixth Sense: Accelerating Organisational Learning with Scenarios* (New York, NY: John Wiley & Sons).

van Wendel de Joode, R. (2005) 'Understanding Open Source Communities: An Organizational Perspective', *Delft University of Technology* (PhD Dissertation), http://repository.tudelft.nl/view/ir/uuid:297bc2ff-956b-436b-addb-98eb1d4a3b4f/, date accessed 11 April 2014.

Vargas, J. A. (2012) 'How an Egyptian Revolution Began on Facebook', *The New York Times,* http://www.nytimes.com/2012/02/19/books/review/how-an-egyptian-revolution-began-on-facebook.html?pagewanted=all&_r=1&, date accessed 11 April 2014.

Varoufakis, Y. (2013) Bitcoin and the Dangerous Fantasy of 'Apolitical' Money, http://yanisvaroufakis.eu/2013/04/22/bitcoin-and-the-dangerous-fantasy-of-apolitical-money/, date accessed 11 April 2014.

von Hippel, E., and von Krogh, G. (2003) 'Open Source Software and the Private-Collective Innovation Model: Issues for Organization Science', *Organization Science,* 14, 209–223.

Walker, B. H., and Salt, D. (2006) *Resilience Thinking: Sustaining Ecosystems and People in a Changing World* (Washington, DC: Island Press).

Walker, B. H., Abel, N., Anderies, J. M., and Ryan, P. (2009) 'Resilience, Adaptability, and Transformability in the Goulburn-Broken Catchment, Australia', *Ecology and Society,* 14(1), 12.

Webster, F. (2006) *Theories of the Information Society* (New York, NY: Routledge).

Wikispeed (2012) WIKISPEED, First Car-Maker in the World to Accept Bitcoin, http://wikispeed.org/2012/07/wikispeed-first-car-maker-in-the-world-to-accept-bitcoin-press-release/, date accessed 11 April 2014.

Wilding, N. (2011) *Exploring Community Resilience in Times of Rapid Change* (Dunfermline: Fiery Spirits Community of Practice).

Wolff, R. (2010) The Keynesian Revival: A Marxian Critique, http://rdwolff.com/content/keynesian-revival-marxian-critique, date accessed 11 April 2014.

Zittrain, J. (2008) *The Future of the Internet: And How to Stop It* (New Haven, CT: Yale University Press).

Žižek, S. (2010) *Living in the End Times* (London: Verso).

DOI: 10.1057/9781137406897.0018

Index

DOI: 10.1057/9781137406897.0019